Human Ecology and
World Development

Human Ecology and World Development

Proceedings of a Symposium organised jointly by the Commonwealth Human Ecology Council and the Huddersfield Polytechnic, held in Huddersfield, Yorkshire, England in April 1973

Edited by

Anthony Vann and Paul Rogers

The Polytechnic, Huddersfield, Yorkshire, England

PLENUM PRESS · LONDON–NEW YORK

Library of Congress Catalog Card Number: 74-1619
ISBN-13: 978-1-4684-2096-8   e-ISBN-13: 978-1-4684-2094-4
DOI: 10.1007/978-1-4684-2094-4

Contributors

Edwin Brooks	Department of Geography and Geology, University of Liverpool, Liverpool
Bruce Dinwiddy	Overseas Development Institute, 10/11 Percy Street, London W1P DJB
Kenneth A Dahlberg	Department of Political Science, Western Michigan University, Kalamazoo, Michigan, USA
Jonathan Holliman	Friends of the Earth, 9 Poland Street, London W1
Angus Hone	Institute of Commonwealth Studies, Queen Elizabeth House, 21 St Giles, Oxford OX1 3LA
Peter Kenyon	54 Devonshire Road, London SE23 3SX
Palmer Newbould	New University of Ulster, Coleraine, Northern Ireland
Hugh Springer	Association of Commonwealth Universities, 36 Gordon Square, London WC1
Anthony Tucker	'The Guardian', 192 Grays Inn Road, London, WC1X 8EY

Foreword
Sir Hugh W. Springer

This book is the outcome of a symposium organized jointly by the Commonwealth Human Ecology Council and the Huddersfield Polytechnic and held in Huddersfield in April 1973. It is the third book to have resulted from the work of the Council and like the other two it illustrates the need for a multidisciplinary approach when examining problems of world development.

Commonwealth countries, like many other countries of the world, are becoming increasingly concerned that their processes of development should be so ordered as to preserve or enhance the quality of human life, and should therefore take account not only of economic considerations but also of all the other factors that must be kept in balance if all man's needs are to be satisfied in due proportion. Human ecology is moving towards a central place in development studies.

The Commonwealth Human Ecology Council (CHEC) was set up to find ways of helping Commonwealth countries to cope with human ecological problems by providing an agency that could promote understanding and pool the knowledge and experience available in this field. Its functions are as follows:

1. To encourage the promotion and assist the growth of organizations concerned with human ecology in the countries of the Commonwealth.
2. To sustain a framework of affiliation through a common council.
3. To stimulate interest at all levels in educative activities and provide a centre and clearing house for information.
4. To encourage, support and co-ordinate studies in human ecology with particular reference to development in the Commonwealth countries.
5. To encourage and support the holding of Commonwealth

seminars and working committees in human ecology with
particular reference to development.
6. To help form among those responsible for decision
   making a proper appreciation of the issues of human
   ecology.

The Council was formally incorporated in 1968, having
developed from an earlier Committee on Preventive and Social
Medicine. Before that, in 1966, a pilot human ecology case
study had been begun in Malta in which the Government and
the University took part. Preliminary reports of that study were
presented to the First Commonwealth Conference on Development and Human Ecology, organized by CHEC jointly with the
University and the Government and held in Malta in the autumn
of 1970.

The proceedings of that meeting, edited by Professor H.
Bowen Jones, were published under the title *Human Ecology in
the Commonwealth,* and included, in addition to the Malta
material, reports from many of the 15 Commonwealth
countries participating in the Conference. Shortly afterwards
the government constituted the Malta Human Environmental
Council which has since become affiliated to CHEC.

Early in 1972 the Second Commonwealth Conference on
Development and Human Ecology took place in Hong Kong,
again organized by CHEC, with the close co-operation of the
University of Hong Kong and with Government assistance; and
studies of Hong Kong by members of that University formed
the chief basis of discussion. Here too one of the main aims was
to encourage the human ecological approach in development
planning, and it is hoped that the report of the conference,
which will soon be published, will be a useful contribution to
this end.

In the field of education, CHEC has concerned itself
primarily with studies of the development of human ecology
courses at tertiary level. Following a preliminary appraisal of
the situation in Australia, a survey of such courses throughout
the Commonwealth is in progress, financed by a grant from the
Commonwealth Foundation. In the closely related field of
information retrieval, CHEC has established an Information
Service. This is designed to serve as a central reference system
through which requests for information on any discipline

integral to human ecology may be channelled to appropriate
sources.

Inevitably, much of the work of CHEC is concerned with the
stimulation of interest in human ecology by personal contact,
but the Council also sees the value of seminars devoted to
specific aspects of the subject. It is in this connection that the
relationship between CHEC and the Huddersfield Polytechnic
has developed which has led to what we hope will be a
continuing series of symposia on aspects of human ecology.

The Polytechnic has a long-standing interest in the multidisciplinary approach to the management of the human
environment and was the first educational institution in the
Commonwealth to become a corporate member of the Council.
The first of the joint symposia was held in March 1972 and was
on tertiary education in human ecology. About ninety
participants from universities, polytechnics, industry and local
and national government, both in the UK and overseas,
attended the symposium and the proceedings were later
published under the title, 'The Education of Human Ecologists'.

The present book is a record of the second symposium, on
'Human Ecology and World Development', which, like its
predecessor, involved contributors from a number of disciplines.
The theme was, of course, wider than that of the previous
symposium and attempted to bridge the gap which undoubtedly
exists between the many people concerned with problems of
the environment and the equally large number whose concern is
with problems of international development.

It generated a valuable debate between ecologists and
economists concerning problems of world development and the
constraints put upon such development by ecological realities.
In this book, the editors have attempted to retain this element
of constructive debate by including summaries of the extensive
discussion that took place.

Introduction

Despite urgent calls for the consideration of environmental problems on a global level, it would appear inevitable that most interest and action should have occurred to date on a domestic, often highly localized level. This may reflect partly traditional areas of concern and partly the levels of organization at which human societies currently respond most rapidly. National schemes for cleaning up rivers and localized schemes of environmental improvement (often carried out with the aid of centralized funds) are areas which have received considerable attention in the United Kingdom.

No doubt time will tell how efficient this type of piece-meal action may be, but it would seem that many of the more serious problems and wider implications of the 'ecological crisis' are receiving comparatively little visible attention. There would appear to be a dichotomy of thinking on such matters—on the one hand some of the more tangible (and easily dealt with) facets of the problem receive attention, while on the other industrialized consumer societies continue to seek yet higher levels of material well-being. It would appear paradoxical that at a time when there is extensive discussion concerning an imminent energy crisis, the domestic economies of Western industrialized nations should be directed towards rapid expansion along fairly traditional lines. However, in addition to the seemingly intractable situation of international competition (some may call it economic warfare), such problems as domestic unemployment and even environmental dereliction may require expanding economic activity as a step in their solution in the current economic climate. There is an implied risk in any case that money which might have been used for development assistance may be diverted towards cleaning up the environment of developed countries.

One hopeful note was struck in 1968, when the United Nations called for a conference on the human environment, at which there could be international discussion of all relevant problems, leading (hopefully) to some multilateral co-operation. The event, which was held in Stockholm in 1972, suffered from a number of severe limitations from the outset. The Eastern European countries and Russia declined to contribute owing to the exclusion of the German Democratic Republic. Ironically, the GDR may soon have full status at the UN. Two further limitations were the exclusion of discussion on the ecological implications of the development, production and deployment of nuclear weapons, and the control of population growth. Although the relevance of the Conference to underdeveloped countries may have diminished somewhat as a result of this latter exclusion, there was some discussion of the priorities for non-industrialized countries in terms of agricultural development, employment, technological development and to a limited extent aid, trade and political participation.

Of far more immediate relevance was the third United Nations Conference on Trade and Development held in Santiago, Chile, also in 1972. While the Stockholm conference may be described as enthusiastic and vague, the Santiago conference has been described as a 'very successful failure'. This was despite the considerable efforts of the 96 underdeveloped countries to present common proposals at the Conference, particularly with regard to the preferential use of new issues of Special Drawing Rights (SDR's) and, more significantly, greater access to the markets of the developed countries for the products of the underdeveloped world.

In the event, the developed world gave few concessions. It was agreed that some specific help should be aimed at the 25 poorest countries in the world. Since none of them is very large, and such selective treatment may break up such unity as exists among underdeveloped countries, this measure may be both economically and politically expedient from the point of view of the developed nations. No doubt with the fluid state of the international monetary system and the enlargement of the European Economic Community, developed nations may have been more concerned with their own national economic affairs. Since any improvement of the human environment in underdeveloped countries appears to depend very largely on a high relative increase in wealth, neither of these Conferences gives any concrete assurance that the developed world intends, at the present time, to substantially assist the underdeveloped world in the achievement of its material aspirations.

If the backgrounds and conclusions of the Stockholm and Santiago Conferences are considered along with such studies as the much debated MIT document, *Limits to Growth,* then the implications for the Third World are as serious as they are inevitable. While some resources have been considered as finite in the past, now all resources are considered in terms of their supplying power, whether capital or renewable. The implications, in the most general terms, for underdeveloped countries, are clear. Even with due allowance made for some measure of population control, and technological innovation, the global resource base may be considered to be finite, in case of capital resources, or limited in supplying power in the case of renewable resources. With the developed countries already dominating the current pattern of use and deployment of global resources, the scope for expansion and development for the Third World must, of necessity, be limited.

There is, of course, the possibility that some underdeveloped countries may profit greatly from the growing demand for resources. This trend is already clearly established in the Middle Eastern oil states and other countries of the OPEC group. We may expect that as the demand for oil and petroleum products continues to increase and the world reserves of oil decrease, there will be an accompanying increase in price. Depending on the scale of increased demand, and the possible discovery of further reserves, we may expect the relative wealth of these states to increase rapidly. It would be nice to think that the demand for other commodities might equally profit more impoverished nations in Africa and Asia in particular, provided that such nations were able to control the influence which western technology had on their endemic cultures. At the present time there are few trends in this direction, but rather more in the reverse. How long this continues to be the case depends on the economic, resource and population statuses of both the developed and underdeveloped worlds.

The symposium on which this book is based sought to examine some of the most important interfaces between the two disciplines of Development Studies and Human Ecology, interfaces which to the present time have received comparatively little academic attention.

In Part I, the present situation of World Development is considered. Peter Kenyon discusses the nature and extent of world poverty, Bruce Dinwiddy describes current and future trends in development assistance, and Angus Hone outlines current and future trends in international trade and finance. Part II is concerned with some of the ecological limits to development. The global ecosystem is described by Professor Palmer Newbould, and Professor Kenneth Dahlberg discusses some ecological effects of current processes of development in less developed countries. Jonathan Holliman describes some ecological approaches to agricultural development, and Anthony Tucker discusses the relationship between development and physical resource utilization. In Part III, where the implications for future world development are discussed, the preceding sections are integrated by a paper by Dr Edwin Brooks. In addition to the formal sessions described, there was every opportunity for economists and ecologists to interchange ideas and viewpoints. Some of these interchanges are published as discussions after each of the formal papers and as workshop and general discussions at the end. They were undoubtedly one of the most important and significant aspects of the symposium.

Acknowledgements

In organizing the symposium and preparing the proceedings for publication we relied heavily on the co-operation of the speakers and other participants and on the help of a number of our colleagues at the Polytechnic. In particular we would like to thank Hazel Taylor, Phyllis Allen, Marjorie Forth, Catherine Hole, Jan Newton-Howes, Shelagh Brooke, Wendy Cole, Findlay Cook, Michael Morphy and Mr F Mickelthwaite of the Polytechnic and Zena Daysh of the Commonwealth Human Ecology Council for help with the organization of the symposium. We would also like to thank Jean Milner and Linda Haley for typing the manuscript and Greenhead Books for arranging a very wide-ranging book exhibition at the symposium. Finally we would like to thank our respective wives, Lynne and Claire, for their helpfulness throughout.

The Polytechnic,   Anthony Vann
Huddersfield,   Paul Rogers
March 1974

Contents

The Nature and Extent of World Poverty

Peter Kenyon

This is an issue which arouses both emotion and apathy. It is a fact of modern life that over half the world's population is undernourished, in ill-health, inadequately housed, illiterate. With population growing, the number of people suffering from absolute deprivation is growing year by year. Although United Nations organizations, such as the Food and Agricultural Organization and the World Health Organization have been confronting governments with these facts, and development and relief organizations such as Oxfam and Christian Aid have been confronting the public with these facts, the problems exist on a greater scale than has ever been experienced in the history of mankind.

Poverty has even been discovered on an extensive scale in the most materially affluent society on earth, the United States of America. It still exists in Western Europe. Although it is less evident than in the developing countries, it is here, it is closer and it is more troublesome. People not in poverty when confronted with the facts of its existence both in our own society and in the world in general react differently. Some believe that those in poverty have no one to blame but themselves. Others accept that something must be done and perhaps acknowledge that they don't really understand why it exists, but contribute money and time to help alleviate it. A few attribute it to the nature of society, declare that it must be changed and then poverty will be overcome. Yet why, in this unprecedented period of general affluence in the industrialized world, should such stark and inhuman conditions be endured by so many?

In order to understand this we have to look at the nature of society and economic activity. Progress has involved casting off the anchor of subsistence, that is, satisfying needs directly. It also imposes needs which go beyond those necessary to keep the body alive. It is only through an understanding of these relationships that we can appreciate the nature and extent of world poverty. I am not going to dwell on the statistical extent of world poverty: most of the data is guesswork. But for those with a predilection for figures I will read a poem by a Malaysian poet included in a recently published anthology called 'Other voices, other places'; it is entitled 'Statistically':

Statistically
it was a rich island
income per capita
one million
per annum

Naturally
it was a shock to hear
half the population
had been carried off
by starvation
Statistically
it was a rich island

A U.N. Delegation
(hurriedly despatched)
discovered however
a smallish island
with a total population
of—2
Both inhabitants
regrettably
not each a millionaire
as we'd presumed
But one the island owner
Income per annum:
Two million
The other
his cook/chauffeur
shoeshine boy/butler
Gardener/retainer
handyman/labourer
field nigger etc. etc.
The very same
recently remaindered
by malnutrition
Statistically
it was a rich island
Income per capita
per annum   C. Rajendra
one million   Malaysia.

Survey of Global Relations

Current world disorder is dominated by a configuration of three phenomena, Science, the Market and the Nation State—The Trinity.

With each scientific advance, new powers have been entrusted to human beings. Science has tended to sift separate pieces of knowledge into discrete categories. Little attention has been paid to the relationships between those categories. Despite this lack of understanding of the relationships between for example engineering and biology, other drives have possessed the utilization of what knowledge we have, the pursuit of material wealth and claims of sovereignty. To quote Francis Bacon, the new gods are 'idols of the market and the idols of the tribe'. Without these the transformation of human activity would probably not have taken place. The consequences of this transformation are three-fold: firstly, the impact on income distribution between individuals; secondly, the distribution of income between private individuals and the public sector; and thirdly, the distribution of income between nations.

During the industrial revolutions in this country, property relations and control of legislative power were such that landowners could remove peasants and cottagers off the land and use new productive techniques in order to increase output and profit. The dispossessed moved into new industrial centres to compete for available jobs and a wage income, leaving the surplus for those in control to either reinvest or spend on lavish consumption.

The problem today is that the surplus in developing countries is insufficient to absorb the dispossessed as wage labour. Similarly, to use a topical issue, the availability of suitable housing to absorb tenants from impounded property is inadequate to prevent increased homelessness. By exchanging goods for money at prices sufficient to ensure a margin between the cost and the sales prices and entrusting these transactions to individuals, the satisfaction of private needs, desires and whims has been more readily achieved than less evidently essential public needs which are not readily saleable. The social evils of industrialization and environmental squalor of urban life were so evidently unacceptable as to lead to protest and violence. Yet it took more than a century before the inability of the Market to correct the private/public imbalance was discredited sufficiently to permit enough public intervention to buy off the 'malcontents' in the satisfaction of indisputable public goods such as health and education. Nevertheless the market has not been sufficiently discredited as yet to do anything to abate the economic boom in the industrialized countries which is threatening the natural fabric of the entire planet. This boom is being fuelled by technological innovation devising new substances and products for insatiable consumers of personal material goods and military hardware. Their demands are fired by the siren call of the good life and a political commitment to growth, full employment and the defence of national sovereignty and superiority.

The nation state provided the framework for internal economic development and the springboard for overseas 'adventure'. Trading led to the establishment of political control and the creation of empires. As industry developed within the nation states of Western Europe the pattern of colonial exchange emerged, capital and manufacturers for raw materials. The same pattern exists today. Although the colonial powers have either been forced out or have astutely withdrawn from express political domination and granted political independence, the economic balance between the former colonial powers and their colonies—the rich and developing—is viciously skewed in favour of the industrialized.

For the new nation states there are no opportunities for 'overseas adventures' and a dwindling stock of resources with which to satisfy their growing internal demands for basic necessities, let alone the 'good life'. As growth in the industrialized countries is pursued relentlessly, so the population of the developing countries is expanding, stretching the time it will take for the most glaring and alarming poverty to be eliminated.

This population explosion is merely a symptom of world poverty. People are living longer, surviving because of the eradication of disease for an existence without much hope. They live in countries which are mostly unable to provide anything more than the most basic of necessities, and in a world community which is largely unwilling to provide what is most needed—a change of priorities. Political recognition of this will not come through exhortation or through our mass communications systems. It will come through a realization that mankind has got to find the means of reconciling his individual desires and the ambitions of nationhood with the wider unities of a shared planet.

Survey of Developments in Poor Countries

It is only in the past 25 years that most of the countries we know as developing have achieved independent nationhood, and a seat at the United Nations, with a check list of problems which would horrify most of our politicians. For the most part their economies are still dependent upon the exchange of raw materials for capital and manufactured goods. Such economic growth that has taken place has been achieved through utilizing their own resources, supplemented in a limited way by financial and technical assistance from the industrialized countries. Although higher rates of economic growth have been achieved than any previously recorded, the potential benefits have been eroded by population growth. The explosion is real enough, but as was suggested above it is a symptom of under-development. The population of this earth will reach 4,000 million within two years, and 5,000 million by 1985. It is understandable that the implications of these projections should be read as mouths to feed. But to suggest, as a recent propaganda poster did, that 'no child need starve when the world plans its family' and by implication that hunger is the result of procreative recklessness on the part of peoples in the poor world indicates a complete lack of understanding of the issues at stake. Many of you are no doubt familiar with the fact that the whole population of the world could stand upright on the Isle of Wight, uncomfortable but no worse than The Underground in the rush hours. Therefore, the main problem is not space, but the capacity of the earth to feed itself.

The physical resources are available. In tropical Africa and tropical America 400 million hectares could be added to the cultivable area. Over much of Monsoon Asia improved irrigation would make two or three crops a year possible instead of the present single crops. This could be achieved without any further technical breakthroughs. The basic problem is priorities: In the mid-sixties a governor of a state in Brazil before the military takeover was reported as saying:

The enormous strip of massape (rich heavy black soil) in the Brazilian North-East is one of the most fertile areas in the world: it is nine times larger than the cultivable land of Japan which feeds 100 million people. But from our land we get only sugar cane and some subsistence products in quantities well below the needs of the 23 million inhabitants of the region. The reason for this is that the exploitation of the soils, where it takes places, is not designed to provide for the needs of the population, but to enrich half a dozen large land owners.

The speaker is now in exile.

So what is the relevance of population control? In most parts of the developing world family planning is regarded with the utmost suspicion—an instrument of maintaining the status quo. This is a view best characterized by de Castro in 1961 when he denounced the industrialized countries,

Which after having looted the world in a manner so shamelessly, so inhuman and so shortsighted that it today realises that the wealth of our planet is being exhausted now admits its bankruptcy and advises the marginal peoples to curb these birthrates so as to save the scraps that remain and leave the exclusive benefits of these to the privileged groups of the moment.

Yet demographic analysis has made it clear that decreases in fertility can contribute materially to growth of per capita income. However, this cannot be achieved quickly. The suggestion that population growth will prevent development altogether can probably be discounted. It is significant that fertility rates are declining in some poor countries, those where levels of education, nutrition and per capita income are higher than the average for poor countries as a whole.

It has been assumed by economists that growth would reduce unemployment and poverty; measures to encourage redistribution have been deliberately shunned in the belief that savings and hence investment and growth would fall. But the experience of growth has been that of itself, it has not led to either a reduction in unemployment or poverty. This is as true in developing countries as it is in the industrialized. This experience and a more considered view of redistribution has led to the conclusion that active measures to provide employment and relieve poverty are more important in actually promoting development and reducing the birthrate. This is not to say that family planning does not have its place. It is just that it will not relieve poverty on its own. Most of those in poverty live on the land. The World Bank estimates that about 1,300 million people in the developing countries are members of farm families. The economies of their nations may be growing. A few farmers may be using those much talked about high yielding varieties of wheat or rice. But for the vast majority these achievements have passed them by. They are not able to pay for irrigation, pesticides or fertilizers. In many areas their land holding is insecure.

The government may have vastly increased its educational budget for urban schools and expensive universities, the town may have developed new industries, jet aircraft may be landing at the new international airports. But these things represent what the majority all too often do not have—rural schools, all weather roads, simple improvements to increase the output of food and provide greater security. In many countries in the developing world the balance between the rural areas, where most people live, and the towns, has been totally disrupted. This is because most development activities have been concentrated in the towns, the gap between incomes in the country and the town has widened and such education as has been provided has encouraged young people to seek their fortunes in the towns. The consequence is urban unemployment. This undue concentration in the towns has many facets—national pride, ease of administration and the growth/industrialization formulae are probably the most important. In addition, the growth of national output, meticulously documented in national accounts statistics, has in many countries not added one job. There is one case quoted by W Arthur Lewis in Kenya where employment fell by 1% between 1954 and 1964. There are two reasons for this: (i) much of the investment that has taken place is wasted on under-utilized prestige projects and has been undertaken with little consideration of the employment that would be created; (ii) machines have been too readily used where jobs could have been provided. The bulldozer, crane and conveyor belt achieve nothing more than men could do equally well. Though this fact may be distasteful to us in this country, labour-saving devices are a luxury poor countries cannot afford in human terms. Perhaps we cannot afford to have them in ecological terms, but I will leave that for others more qualified than I to take up.

It is not growth *per se* which is at fault, it is the way in which resources are utilized in an ill-considered pursuit of growth.

That is the responsibility of politicians and experts who either administer resources directly or create the framework within which they are utilized. Development, that is a growth in production activity that improves the quality of life, is an extremely complex process. It requires a balance between town and country, between employment-creating and employment-destroying activity, utilization of resources and destruction of resources.

There is now widespread agreement that if broad based development is to take place, policies to improve and alter social conditions are of fundamental importance. In the developing countries, this means rural development; agriculture, then light, labour intensive industry, and lastly, heavy industry. Programmes that deal with only one aspect or another handicap the prospect of achieving those objectives. An integrated social and economic approach is required, which demands courageous political leadership and support. The only country where this seems to have been achieved is China. Its entry into world-wide institutions is an important step both for developing and developed countries. From what little has been heard already we all have much to learn from that country.

Whilst we are quite prepared to accept the need for radical change in the poor countries, there is considerable reluctance to admit there is anything fundamentally wrong with our own values and institutions. Yet is is the persistence of poverty in the affluent industrialized countries, the imbalance between private consumption and public squalor, that points most readily to the inadequacy and deficiencies of existing institutions and values.

The US President's Commission on Income Maintenance Programs (1969) concluded, 'Our economic and social structure virtually guarantees poverty for millions of Americans. Unemployment and under-employment are basic facts of American life'. Yet it goes on, 'Society must AID them or they will remain poor'. I have not been able to find such a frank disclosure in any comparable Royal Commission Report. But it is the familiar tone of that conclusive remark which is most illuminating. It follows exactly the pattern of thinking which has conditioned the response of the Western World to the facts of world poverty.

The Western approach to development has a beguiling

quality—a gloss of humanitarianism. Being aware of the condition of two-thirds of the world's population and inclined to gestures of commiseration it is our collective wish that they will succeed in achieving the material prosperity we currently enjoy and so we AID them.

It is not humanity which is misplaced, but the expression of it in a context which is not understood. To quote Gunnar Myrdal in *Asian Drama,* 'It is the ethos of scientific enquiry, that illusions, including those inspired by charity and goodwill, are always damaging.' A trap which Myrdal himself possibly fell into is presuming that capitalist production could facilitate development in South Asia. In a review of Professor Myrdal's vast study Paul Mattick concludes:

It does not occur to him (Myrdal) that it may be far too late to have any kind of progressive development under the auspices of capital production, and that the development of South Asia, most of all, might depend on a further non-capitalistic development of the advanced nations.

The sheer force of value transference is the most complex developmental problem with which we have to contend. As more and more studies of the extent of poverty both in the industrialized and developing world come to the same conclusions—the need for social change—the need for a change in basic relations between men and between nations, so the need for change in values and priorities becomes more acute. Yet our political response is conspicuous by its absence. It is accepted that the system of property relations in traditional agricultural society will not provide for the current needs for developing countries. But it is less readily accepted that the system of property relations in modern industrial society is equally incapable of providing for our current and future needs and aspirations.

Poverty, the extent of which there is no need to encapsulate in figures and statistics, will persist unless there is a basic change in the system of rewards and penalties influencing productive activity, and the institutions through which it is achieved. It will require a political and personal commitment far more courageous than that which fired civilization after World War II. The domestic political risks of tearing away the gloss of material affluence are seemingly too great.

Perhaps we should start by examining the nature of our development and ask ourselves whether the quality of life has been vastly improved, why it is we have to take longer holidays away from work, whether we need all the goods we consume and what services they really provide us with, and whether there are less wasteful and stressful ways to achieve the things we do achieve. We should stop pretending that what we have achieved is superior to anything else and that our life is the 'good life'. We should accept that we are just as much part of the developing world as the poorest.

With greater knowledge and collective responsibility the survival of man and the end to his poverty are more likely to be achieved.

Discussion

JOHN TYM

It is surprising and disheartening that leaders of less developed countries do not make fundamental criticisms of the way we in developed countries use our resources. If only the leaders of India, or Pakistan or perhaps one of the African countries would point out, for example, the amount of money being wasted on projects such as Concorde, where every plane will use each year, fuel to a value equal to a substantial part of their development expenditure. We now learn that the Chinese are contemplating buying Concorde, although in view of the nature of the project they may be doing this with a view to ruining our economy. But joking apart, how can such an expensive project be justified in such a country?

PETER KENYON

There are two points here. Firstly, why is such criticism not forthcoming? The answer is that such criticism would be pointless, as it would be ignored by the countries at which it is directed. The less developed countries which are currently attempting to pursue policies of self-reliance are those which face increasing isolation by the developed countries. The nature of the world political system is such that criticism of the developed countries would just make this isolation more severe. Also, such criticism assumes a dependency on the developed countries by any less developed countries making that criticism. Whether such dependency exists, it is a fact that a number of less developed countries are trying increasingly to exercise their independence.

With regard to Concorde, China does not have an aircraft industry of its own, its aircraft are imported. Presumably on the Chinese ordering of priorities, they consider that it is desirable to have an aircraft which can carry people around the world quickly, and presumably this is a means of overcoming the isolation from the rest of the world currently experienced by the Chinese.

G S PURI

I think developing countries are conscious of the wasteful use of resources by the developed countries, but being poor and undeveloped they would be snubbed if they protested openly. I feel that we have to be very careful about defining poverty and wealth. They cannot be measured simply in terms of GNP or monetary income, because although in such terms western countries are rich, in many other ways they experience poverty.

Take as an example the side-effects of the industrial development which has brought these countries wealth. Can we really say that the levels of pollution experienced by so many industrial nations are manifestations of wealth? If we were really effective in clearing up the mess left by industrial development, then our monetary wealth would be much lower than at present, and in some cases developed countries might end up poorer than the developing world.

Another major factor is food and nutrition. Malnutrition due to lack of particular kinds of food is common in poor countries. But malnutrition and dietary disorders due to too much food are common in western countries. Is that wealth? Diseases of many kinds are common in poor countries and cause much suffering and a high mortality rate. But in western society we are so dependent on pills that as much as half of the total population will be on pills at any one time. Is that health? Too many hospitals does not necessarily mean a healthy population.

With these things in mind I would suggest we question carefully our basic ideas about wealth and poverty. The question 'what is poverty?' is a very difficult one to define at present and some thought should be given to the ecological context to this factor.

PETER KENYON

I completely agree with you on those comments. In a sense we are all part of a developing world, especially regarding the maldistribution of factors affecting human well-being in countries such as this. Of course, the almost total lack of monetary resources in many parts of the third world does result in human misery. This aspect of poverty cannot be overlooked.

L J HALE
I feel you rather underestimated the importance of the growth of population as a factor hindering world development. I am not good at remembering figures but I think that at the present time the world's population is growing annually by a figure equal to the total population of Britain. This must impose severe pressures on development prospects, not least the basic problem of producing enough food for this expanding population.

Another point I would like to raise concerns development assistance. One appreciates that there is a need for a greatly increased level of financial aid to less developed countries, but are we not likely to witness problems in assimilating and using this assistance effectively? Just what capabilities do these countries have to utilize increased financial assistance effectively?

PETER KENYON
I accept that the need to prevent excessive population growth is essential but must emphasize that population control is only one part of a total requirement for development. Time and again, it has been shown that population control measures introduced in areas of malnutrition, illiteracy and poverty have been of very limited value. Such programmes are seldom effective unless efforts are made to overcome these problems. The population question is part of the totality of world poverty, not a separate issue, and control programmes will only be effective to the extent necessary when forming part of general development programmes.

With respect to your point concerning levels of aid and the ability of countries to absorb aid, I think that the increase of levels of aid is an important means of stimulating economic development, but is only one means. As a whole I think that other methods of stimulating economic activity are preferable to aid and have no doubt that Angus Hone will be considering these in some detail in his paper. Regarding aid, I do not see any difficulties on the part of the less developed countries in absorbing increased levels of aid from the developed countries.

ANGUS HONE
If I could just put this in perspective, if someone suggested increasing trade with a less developed country to the order of say $30 m and then worried about whether that country could absorb such increased export revenue, one would be considered to be worrying needlessly. But if aid were to be increased to that country by say, $7 m then people would immediately start worrying about absorptive capacity. Generally, levels of aid are about one-tenth of the amount of income earned by less developed countries for their exports. They are able to handle such income and use it effectively to import goods, why then do we place such emphasis on absorptive capacity of aid? 'Absorptive capacity' is one of the most 'smelly herrings' in the whole business of international development.

KENNETH BARLOW
I would like to take issue with the speaker on the question of integrating population programmes with development programmes. A recent report in *The Lancet* suggested that an immediate effect of improving nutritional levels is to increase the birth rate.

PETER KENYON
I would like to see that paper as it would run counter to a number of reports to the contrary, indicating that unless integrated into development programmes, population control programmes on their own are likely to prove ineffective.

SARAH WELLS
There are dangers in discussing population problems on a national or international basis when in reality they are problems best approached at the family level. It is important to recognize that in countries with little or no social security systems, children provide a means of support for parents in later life. The decision on the part of a husband and wife as to the number of children to have will depend very much on this kind of factor. There are reports suggesting that a mother whose first child dies in infancy will end up having more children than a mother whose first child lives. In view of this kind of information, I would be interested to know whether *The Lancet* report mentioned earlier was a result of short-term or long-term studies.

PETER KENYON
This kind of information, suggesting that, perhaps paradoxically, families will end up being larger in areas with a high infant mortality rate, is of the sort which supports the idea that population control programmes, though important, should be seen in the context of broadly-based development programmes.

Current and Future Trends in Development Assistance

Bruce Dinwiddy

Preface

The term 'development assistance' is often given an over-narrow interpretation. It makes people think merely of official aid programmes, and perhaps of the development work of the private voluntary agencies, whereas it really embraces any action taken by rich countries, individually or collectively, which is prompted primarily by a desire to promote overseas development rather than by immediate rich country self-interest.

One only needs to look at the agenda for the 1972 UNCTAD conference, which was drawn up largely at the behest of developing countries themselves, to appreciate how broad a range of rich country policies have some bearing on the economic or social welfare of poor countries [1]. Quite apart from direct aid policy, they include, for example, the attitudes of rich countries towards the conduct and regulation of international trade and the reform of the international monetary system, as well as to the transfer of technology and the control of multinational corporations. There is a correspondingly wide range of areas in which the governments of rich countries, by actively influencing market forces in favour of developing countries, can engage in development assistance. Nevertheless, in order to avoid overlap with the subject-matter of other papers presented to the symposium, this paper is limited to a discussion of trends in development assistance in the narrow sense of financial aid and technical assistance. Moreover, except in passing, it does not discuss flows of private commercial capital (including export credits). Although the latter may contribute to development, their volume and their geographical and sectoral distribution are principally determined by commercial considerations, divorced from any active concern to assist developing countries.

HEWD–2

The Volume of Development Assistance

For purposes of aggregation or comparison the volume of development assistance is usually measured in international currencies—most commonly in dollars. On the other hand, no-one pretends that, dollar for dollar, official aid provided by different donors (or even by the same donor in different circumstances) always has the same value in terms of either its cost to the donor or its contribution to development. The subject of aid quality is taken up in the later sections of this paper.

Attention has been particularly focused on aid volume as a result of the way in which aid was treated at the early conferences of UNCTAD (1964 and 1968) and, more recently, in the International Strategy adopted by the United Nations General Assembly (October, 1970) for the Second Development Decade. The UN Strategy sets as a target—which a few donors, notably the USA and the UK, have explicitly declined to accept—that: 'Each economically advanced country will progressively increase its official development assistance to the developing countries and will exert its best efforts to reach a minimum of 0.7% of its gross national product at market prices by the middle of the Decade'.

It is currently an internationally agreed convention that, for the purposes of relating an aid donor's performance in relation to the 0.7% target, aid should be calculated net of amortization (i.e. capital repayments) on past loans. Interest payments are not taken into account; and although a true picture of the net flow of official aid resources during any one period could only be gained by deducting interest as well as amortization, this paper adopts the conventional method of presentation.

Table 1 shows that, in 1971 the 16 member countries of OECD's Development Assistance Committee (DAC), taken together, were exactly halfway to reaching 0.7%. Although the total volume of aid increased between 1960 and 1971 from $4,665 m to $7,718 m, a large part of this increase was offset by inflation; and, since DAC countries' gross national product was expanding, the 1971 total, in proportion to combined DAC GNP, was barely two-thirds of the level reached in 1960 (at the beginning of the First Development Decade, and before any aid target had even been adopted [2, 3]). This 1971 total did, at

Table 1 Net Flow of Official Development Assistance to Developing Countries and Multilateral Organizations, 1965 and 1970-71*

	Value ($m)				As a % of GNP†				Ranking on GNP share			
	1960	1965	1970	1971	1960	1965	1970	1971	1960	1965	1970	1971
Australia	59	119	202	202	0.38	0.52	0.59	0.52	6	4	4	4
Austria	‡	32	19	10	‡	0.34	0.13	0.06	‡	9	16	16
Belgium	101	102	120	146	0.88	0.59	0.48	0.50	3	2=	5	5
Canada	75	97	346	340	0.19	0.19	0.42	0.37	11	11=	6	9
Denmark	5	13	59	74	0.09	0.13	0.38	0.43	13	14	7	7
France	823	752	971	1,088	1.38	0.75	0.66	0.67	2	1	1	2
Germany	223	456	599	734	0.31	0.40	0.32	0.34	7=	7	10=	10
Italy	77	60	147	183	0.22	0.10	0.16	0.18	10	15	14	14
Japan	105	244	458	511	0.24	0.28	0.23	0.23	9	10	13	13
Netherlands	35	70	196	216	0.31	0.36	0.63	0.60	7=	8	3	3
Norway	5	11	37	42	0.11	0.16	0.32	0.33	12	13	10=	11
Portugal	37	22	41	99	1.45	0.59	0.64	1.42	1	2=	2	1
Sweden	7	38	117	159	0.05	0.19	0.36	0.45	14	11=	9	6
Switzerland	4	12	30	28	0.04	0.08	0.14	0.11	15	16	15	15
UK	407	472	447	561	0.56	0.47	0.37	0.41	4	6	8	8
US	2,702	3,418	3,050	3,324	0.53	0.49	0.31	0.32	5	5	12	12
DAC Total	4,665	5,916	6,840	7,718	0.52	0.44	0.34	0.35				

Notes: * Net of amortization but not of interest; at current prices. † At market prices. ‡ Not available.
Source: DAC Review 1971, Statistical Tables, 2, 9. DAC Review 1972, Statistical Tables, 2, 9.

least, represent a small improvement on 1970: the DAC Secretariat estimates that in real terms (at 1970 prices and exchange rates) total aid volume increased by about 5%. But in relation to GNP, and with the exception of a similar small improvement in 1967 (0.42% compared to 0.41% in 1966) overall DAC performance deteriorated every year throughout the 1960s.

During this time, some countries have done much better—or worse—than others. For example, the performance of the United States (in per caput terms, still the richest DAC member) has been particularly disappointing. And, since the USA contributed such a high proportion (58%) of the 1960 total, the US decline has heavily influenced the figures for DAC as a whole: US aid as a percentage of GNP dropped from 0.53% in 1960 to 0.32% in 1971, while the DAC average dropped from 0.51% to 0.35%. For the United Kingdom, the corresponding percentages were 0.56% and 0.41%.

The performance of other long-established donors—notably Belgium, France, Portugal (almost all of whose aid is concentrated in her three African provinces) and the United Kingdom—has also tended to decline, though Portugese flows increased sharply in 1971 and France has dropped only just below the 0.7% target. Several newer donors, on the other hand, have been fast expanding their programmes: Canada, Denmark, Netherlands, Norway and Sweden have all achieved substantial increases both in absolute (money) terms and in relation to GNP. Japan's programme has also increased—nearly fivefold in dollar terms. But, mainly as a result of the country's phenomenal rate of growth, Japan's performance in relation to GNP is virtually unchanged.

Looking to the future, it is clear, given the enormous wealth of the USA relative to other donors (and notwithstanding the prospective continued growth of Japan), that during the 1970s overall official development assistance levels will continue to be heavily weighted by the individual US performance, and because of the length of time required in most countries to translate legislative authorizations first into commitments and then into disbursements, it is already possible to project, reasonably accurately, the individual and collective performance of DAC countries in the target-year (1975).

Table 2 Projected Flow of Official Development Assistance Measured as a % of Gross National Product*

	1972	1973	1974	1975
Australia	0.59	0.59	0.59	0.60
Austria	0.17	0.19	0.22	0.25
Belgium	0.54	0.58	0.62	0.66
Canada	0.48	0.51	0.55	0.59
Denmark	0.48	0.53	0.58	0.64
France	0.65	0.65	0.65	0.65
Germany	0.33	0.36	0.36	0.38
Italy	0.16	0.16	0.16	0.16
Japan	0.28	0.32	0.36	0.40
Netherlands	0.70	0.74	0.76	0.78
Norway	0.47	0.56	0.67	0.75
Portugal	0.45	0.45	0.45	0.45
Sweden	0.50	0.56	0.65	0.71
Switzerland	0.22	0.26	0.30	0.32
UK	0.41	0.41	0.45	0.46
USA	0.30	0.28	0.26	0.24
Total	0.36	0.36	0.36	0.37

* Based on World Bank estimates of growth of GNP, on information on budget appropriations for aid, and on aid policy statements made by governments.

The information shown in Table 2 was published by the World Bank (as an annex to the President's Annual Address to the Board of Governors) in September 1972. The Netherlands, Norway and Sweden are all expected to attain the 0.7% of GNP by 1975. Substantial increases are also anticipated from Belgium, Canada, Denmark and Japan; but US development assistance may meanwhile drop to only 0.24% of GNP, and the overall DAC average is expected to show only a very small improvement, to 0.37% of GNP. This is only just over half of the 0.7% target; and as yet there is no indication of further significant improvements during the second half of the decade.

The immediate outlook is thus not encouraging, though it should be added that, without the watchdog function fulfilled by DAC through its regular review and comparison of the performance of individual member countries, the prospects even of maintaining present standards of aid performance (relative to GNP) might be considerably worse.

Paul and Ann Ehrlich, on the other hand, in *Population*,

Resources, Environment, their remarkable over-view of major issues in human ecology, have called for really dramatic measures to redistribute the world's wealth, such as would make current aid levels look ridiculously small; and they quote the separate proposals, by Britain's Lord Snow and Russia's Andrei Sakharov, that all rich countries should mount a massive international development assistance effort involving some 20% of rich-country GNP over a 15-20 year period [4]. Even if it was backed by the necessary political will, aid on this scale would require fundamental restructuring of rich-country economies; and, however genuinely altruistic rich countries' intentions might be, their motives might be suspected by developing country governments wary of cultural neo-colonialism. In any case, we must admit that the first essential, a new interpretation within developed countries of the interdependence of the world's rich and the world's poor, does not yet exist; and it is more realistic, in the remainder of this paper, to concentrate on development assistance trends which, however inadequate they be, are practical possibilities in the foreseeable future.

We shall also be mainly concerned with trends in *official* development assistance. First, however, we should recognize the relatively small, but important, role of private voluntary agencies. Relevant statistics (merely for DAC countries) have only been available since 1970; but in both 1970 and 1971 the aid and relief expenditures of these non-profit-making organizations, all in grant form, amounted to $855 m and $890 m respectively. This corresponded in each year to 0.4% of total DAC GNP, and was therefore equivalent to slightly more than 10% of the total official assistance extended during the same two years. There is no information on which to base future projections; but the current scale of contributions from individual DAC members varies much more widely than their performance on official aid. In particular, although in both years the largest contribution in terms of GNP was made by Sweden (0.07% in 1971), the USA (0.06% in both years) is well above the DAC average and therefore stands far ahead of any other country in terms of absolute volume (with $588 m, 66% of the DAC total, in 1971). By comparison, the 1971 contribution of UK voluntary agencies was only $34 m, 0.02% of GNP.

Aid Quality

From the donor point of view, the most obvious aspect of aid quality (and also the most easily evaluated) relates to financial terms. Aid terms, with respect both to individual loans and to the whole aid programmes, are assessed by calculating their 'grant element' [5]. The DAC, through its regular appraisal and comparison of different national aid programmes, has played an important part in encouraging donors to 'soften' their aid; and DAC's latest 'Terms Recommendation', adopted by DAC members in October 1972 at their annual High-Level Meeting, calls on each member to achieve an overall (weighted) grant element, for all its aid commitments in a given year, of at least 84% of their combined nominal value. The new Recommendation also enjoins members to co-operate in harmonizing the terms on which aid is extended to individual countries, and it sets a special target, representing an even greater degree of concessionality, for the terms of aid to the 25 'least developed' countries (as identified in 1971 by the UN General Assembly).

Except for Japan, all major donors are already virtually achieving, or even exceeding, the 1972 overall target. Therefore, although some donors may now soften the terms of aid extended to especially poor or needy countries, the new Recommendation will be more important in keeping donors up to their present mark than in stirring them towards more generous initiatives.

An obvious shortcoming of the established paraphernalia of aid targets, performance rating and so on, is that, although they purport to provide some indication of donor effort, they are not directly related to aid's results in terms of recipients' development. The DAC does stipulate that official development assistance should only include contributions 'administered with the promotion of economic development as the main objective'—thus excluding, for example, military aid—but it is no secret that the scale and the distribution of most aid programmes, particularly those of the longer-established donors, are also heavily influenced by strategic, political and commercial factors.

In practice, a growing proportion of official development assistance is partially immunized from these considerations of predominantly national self-interest by being channelled through multilateral agencies. (The immunity is not complete:

it is sometimes suggested, for example, that the UK's readiness to support the World Bank's soft-lending affiliate, the International Development Association (IDA), partly reflects an awareness that—when all the major donors subscribe to a 'replenishment' of IDA resources—a disproportionate amount of the resulting IDA loan money comes back to the UK in payment for British goods and services.) The proportion of total DAC official development assistance taking the form of contributions to multilateral organizations approximately doubled during the 1960s. But even in 1971 it amounted to only 17% ($1,287 m) of the total, and the great majority of aid is still given bilaterally. This discussion of aid quality is therefore largely centred on bilateral aid (though some of the points apply to multilateral aid as well). The subject of multilateral aid is treated, separately, in the final section.

The effective quality of aid, in terms of its results, depends on the policies and dispositions of recipients as well as of donors. (It should also be remembered that, no matter how wise their policies, some of the countries which most urgently need aid are among the least likely to be able to use it to achieve early and spectacular results.) At the same time, there are two outstanding characteristics of existing aid programmes, both the exclusive responsibility of donors, which severely restrict aid's potential contribution to development. These are, respectively, the present pattern of geographical distribution and the practice whereby the majority of aid is tied to purchase of goods or services supplied by the donor.

The distribution of official aid extended by each of the large donors, including Britain, is biased by a range of historical, political and commercial factors—none of them readily quantifiable. For the most part, however, despite the efforts of the World Bank to improve the co-ordination of different bilateral and multilateral programmes to particular countries, donors make little conscious attempt to direct their aid to countries which are receiving relatively little aid from other sources. The resulting overall distribution cannot be said to correspond to any conceivable criterion of need—as may be judged from the fact that on a *per caput* basis four of the largest recipients over the period 1969-71 were Malta ($45 *per caput* pa), Gabon ($42), Jordan ($27) and Tunisia ($25), while the annual *per caput* aid receipts of both India and Nepal, during the same period, were only $1.72. At present, countries with *per caput* GNP of less than $150, together comprising some 60% of the developing world's population, receive only just over half of DAC's total official development assistance.

There is a glimmer of encouragement in the fact that the 1972 UNCTAD conference unanimously recommended a wide range of special measures—and in particular more aid, on appropriately soft terms—to assist the 25 'least developed' countries; and to the extent that some donors redistribute some of their aid in favour of these countries, making corresponding reductions in the rate of their commitments elsewhere, this may be desirable on grounds of equity. But it should be noted that the total population of these 25 countries is only 150 m, compared to a total population, in all developing countries, approaching 3,000 m (including 700 m in the Asian sub-continent alone).

More generally, there seems to be little immediate prospect of any substantial changes in aid's present allocation—either by DAC as a whole or by Britain. Indeed, a completely rational distribution of aid, on purely developmental grounds, is perhaps an unattainable ideal. Nevertheless, two obvious points may usefully be reiterated: there is a need for increased allocations to very large, very poor countries, notably India and Bangladesh; and there is a need to exclude, at least from so-called official development assistance, the very high levels of aid administered to relatively rich countries (as, in the British case, to Malta), where the donor's *main* objective is clearly not to promote development. In the medium term, the growing share of newer donors—less biased by non-developmental considerations—may serve to accelerate at least some measure of redistribution to more needy countries: whereas longer-established donors still tend to be bound by historical links or (in the US case) by the requirements of Cold War diplomacy, the Scandinavian countries, in particular are notable for a much more resolutely altruistic aid philosophy. Goran Ohlin writes, of Swedish development assistance, that: 'Aid has never been seen as an enterprise in which Sweden herself could have any legitimate self-interest beyond that of her stake in global peace and survival' [6].

The other important constraint to aid's potential effectiveness is the practice of tying. Almost all donor countries restrict a substantial amount of their aid to purchase of their own goods and services. Much of this tying is essentially informal (on the mutually understood basis of 'If you don't buy from us this time, we might not give to you next time') and it is therefore not possible to calculate exactly what proportions of aid are tied; but the UK Government, for example, officially estimates that, leaving aside the cost of technical assistance (which is almost completely tied) and certain financial aid categories (including pensions for former colonial servants) to which the tied-untied distinction is not strictly applicable, a total of 64% of UK bilateral aid was effectively tied in both 1970 and 1971 [7].

The main purpose of tying is to protect donors' balance of payments. Its principal disadvantage is that it reduces the value of aid to recipients: by raising the cost of imports which might be obtained more cheaply from other countries; by imposing additional planning and administrative problems; and, most seriously, by biasing development expenditures towards programmes and projects having a relatively high direct import content. This last consideration is particularly important, in the face of growing unemployment in many developing countries, when there is increasing awareness of the need for more aid to support the local costs of rural development expenditures (including the promotion of more appropriate, relatively labour-intensive, technologies), of better oriented education and training, and of family planning programmes.

DAC members generally accept that aid would be more useful if it were untied; and they know that, in so far as developing countries characteristically do not stock-pile reserves of foreign exchange, untying would be unlikely to damage their *collective* balance of payments position since untied aid money would in any case revert to them in due course. Nevertheless (understandably), for fear of losing export markets, none of the major donors is prepared to untie its aid unilaterally—i.e. unless other donors do the same. It is therefore much to be regretted that mutual discussions on untying, which were taking place under DAC auspices but were discontinued in August 1971 when President Nixon introduced his New Economic Policy, have still not been resumed. These talks were immediately concerned with untying of contributions to multilateral agencies (most of which are untied already) and with the untying of aid loans. The tying of bilateral grants (which account for more than half the bilateral aid which might be untied) would thus continue even if these DAC discussions were successfully concluded; but untying of loans would be a significant step—not only because it is particularly important that the development contribution of loans (repayable at their face value) should not be impaired, but also in establishing a principle which might subsequently be extended more widely. Unfortunately, further progress—even on loans— is unlikely in the present context of world trade and monetary negotiations; but even if the USA continues to reserve its position, there is no reason why other countries should not join in more limited talks. At the very least, for example, EEC countries could all agree that their bilateral loans could be spent on goods and services supplied by other EEC members.

Apart from the policies of donors, the effective quality of aid also crucially depends on true commitment, on the part of recipients, to wise development priorities. Some recipients are also relatively inefficient administratively, or are lacking in vital technical skills. But whereas these latter shortcomings can be partly bridged with the help of appropriate technical assistance, some countries, by virtue of their development strategies, may clearly seem to be more deserving than others. This is already a factor determining geographical distribution. At the same time, it is virtually impossible for a donor—when comparing different recipient strategies—to take an objective view: Western countries (and the advisers they provide under technical assistance) are inevitably influenced by various prejudices in favour of (or sometimes in reaction to) particular institutions, technologies, and economic values. Given, however, that disparities in individual wealth tend to be much greater in poor countries than in rich countries, and that it has been seen (e.g. in West Pakistan and Brazil) that rapid economic growth does not necessarily in itself lead to really widespread improvements in living standards, it is a reasonable generalization that more aid should be devoted to countries explicitly endeavouring to improve the conditions of life of the poorest of their populations. In many countries, this would imply a substantial redistribution of effort and benefits in favour of the rural areas, since this is where most of their peoples live; and extensive (and politically painful) land reform may sometimes be an essential pre-requisite to significant progress.

Multilateral Aid: A Current Vogue or a Future Trend?

The intricate structure of the contemporary aid framework—still mainly consisting of bilateral links but more complicated than it used to be since most recipients now receive aid from many different sources—is largely a historical accident. And, to anyone starting from the premise that aid is essentially a moral obligation, incumbent on all rich countries, whose only self-interest in giving it is their common stake in the world's future, the present system must seem both inefficient and anachronistic.

Ideally, it might appear much more sensible for all aid to be given multilaterally by specialized institutions, able to take a relatively impartial view of the needs of different recipients and in a better position (in co-operation with recipient governments) to ensure the proper co-ordination of individual country aid programmes. On the other hand, although as we have already noted the share of multilateral aid in total official development assistance had risen by 1971 to 17%, there are at least three reasons why this utopian state of affairs will not be realized in the forseeable future. First (and least important in practical terms), a major switch to multilateral aid would not be welcome to some recipients: even if remaining colonies and dependencies continued to receive aid bilaterally, many independent countries—mainly the smaller ones or those most recently independent—would almost certainly receive a smaller share of the total cake if all aid was apportioned multilaterally. Second, several major donors would not at present be willing to contribute to multilateral institutions a volume of resources equivalent to that which they currently disburse bilaterally. Some of the newer donors, or those without close historical links with particular parts of the developing world, might support such a change: in 1971 Denmark (51%), Norway (57%) and Sweden (57%) all already disbursed more than half their aid in the form of multilateral contributions, and a recent report on US aid [8] recommended that the USA (17% in 1971) should increase its multilateral contributions so long as other donors did the same (thereby counter-balancing the US weight in the multilateral system). Against this, France (12%) and the UK (13%) are both at present very unlikely to give multilateral aid significantly increased support.

The most crucial bar to increasing multilateral aid—and this partly explains the reluctance of some donors (especially those who have built up their own specialized aid machinery), to switch towards it—is that the established system does not offer an obviously more efficient channel for aid disbursement. The multilateral institutions include wholly global bodies (FAO, ILO and other UN specialized agencies), the Western-dominated World Bank Group, the regional development banks, and the EEC's European Development Fund—all enjoying considerable autonomy which they would be reluctant to abdicate in favour of a single agency. Any large expansion of multilateral aid, if it was not preceded by more radical changes involving amalgamation of some existing agencies, would need to be built on closer co-operation between these institutions. Also it would be particularly important to define more clearly the relationship between the World Bank, Washington-based and now having a fairly well-established role in promoting better co-ordination of capital aid, and the United Nations Development Programme (UNDP), with its widely distributed field administration and whose new system of country programming is facilitating a more rational distribution of technical assistance through the specialized agencies. Some people, of course, would like to see the whole international machinery of aid administration done away with altogether. It would be quite possible to give aid *ex gratia,* leaving recipients completely free to choose how it should be spent; and one attraction of distributing aid in the form of IMF Special Drawing Rights (SDRs) is that 'the link' (between aid and the creation of international liquidity) could operate in just this way. But it would be hard to arrive at a criterion for distribution which did not involve *some* subjective judgment as to which countries deserved how much; and, again, donors, whatever their aid objectives, would not yet be prepared to resign their rights of supervision and at the same time maintain aid disbursements at anything approaching their current levels.

It is more probable that the rest of the 1970s will witness a gradual expansion of aid, roughly conforming to current patterns—i.e. with bilateral aid continuing to predominate—but with newer donors taking a greater share of the total. Any marked expansion of multilateral aid is more likely to be channelled through regional institutions rather than through the World Bank and the UNDP. In particular, we may see the development or consolidation of a number of distinct north-south axes—between the USA and Latin America, the enlarged EEC and Africa, and Japan and South-East Asia. The position of the USSR and China remains uncertain; and we must perhaps accept that the mounting of a truly *global* development assistance effort would presuppose a further improvement in political relations between the Great Powers, and a new feeling of moral responsibility—among all rich countries—towards the world's poor.

References and Notes

1. *United Nations Conference on Trade and Development.* Cmnd 5134, HMSO December 1972.
2. DAC countries' total *gross* flows in 1971 were $8,822 m: i.e. capital repayments amounted to $1,104 m. DAC interest receipts were approximately £520 m, so that the 'true' net flow of official aid resources was about $7,200 m.
3. Official development assistance totalling about $28 m was also contributed in 1971 by New Zealand and Finland, neither of whom belongs to DAC. More significantly, the centrally planned economies are thought to have extended total economic aid of about $1,800 m (the greater part disbursed to Communist developing countries), including $1,100 m from USSR, just over $300 m from East European countries and, very approximately, $400 m from China; but little more is known, either about Communist aid in 1971 or about future trends.
4. See P R and A E Ehrlich, *Population, Resources, Environment,* W H Freeman, San Francisco, 1972.
5. A loan's 'grant element' is equivalent to its nominal recorded value *less* the discounted present value of the required amortization and interest payments (using a 10% discount rate).
6. G Ohlin, 'Swedish Aid Performance and Development Policy', *Review 6,* Overseas Development Institute, to be published in May 1973.
7. *An Account of the British Aid Programme,* HMSO, 1972, p. 8.
8. *US Foreign Assistance in the 1970s: a New Approach,* Report to the President from the Task Force on International Development, Washington DC, March 1970.

Discussion

A FORWARD

Could I ask Mr Dinwiddy just how useful he thinks tied aid is? I recently spent three years in a developing country, Ethiopia, and witnessed what seemed to me to be a very bad example of wastage of money through the tying of an aid project. The project concerned was at the Haille Sellassie I University in Addis Ababa, an expensive project in which 40% of the construction costs had to be in the form of materials imported from the United States and if one adds to that the consultants' fees and the like, one finds that over 50% of the aid finance was spent in the United States.

Ethiopia will be repaying the loan until well into the 21st century. My strong impression, as an architect, was that the whole project could have been built considerably more economically and more efficiently, had not the loan been tied in this way. I wonder whether you could comment on this specific instance and also on the wider context in the desirability of tied aid.

BRUCE DINWIDDY

I am sure your comments on the particular project are fair. There is an important distinction to be made in the matter of aid tying depending on whether the aid is given as a grant or whether it is in the form of a loan. If the former, then it is, perhaps, more easy to condone the tying of that grant, or at least a portion of it, to the donor country, although even then there is little doubt that the efficiency of the aid is likely to be marred. The tying of aid dispensed as a loan is much more open to argument and the example you quote is typical of many, showing clearly the disadvantages of such tying to the recipient.

The discussions which took place on tying under the auspices of the Development Assistance Committee (DAC) and which might, one day, be resumed, were concerned specifically with the untying of aid loans. It seems that we have to accept this as a first step although I fully agree that the aim should be the general untying of all aid.

SARAH WELLS
On the question of the tying of aid and the whole value of aid we have a double difficulty arising from the opinions of different sectors of the electorate of developed countries such as Britain. First, there are those who say that aid itself is of virtually no value to the less developed countries particularly when heavily tied. Whether they are right or not, they do give support to the majority who are generally against any help to less developed countries. Secondly, there are those who consider that aid is desirable but that it should give some benefit to the donor and therefore support tying. These groups are in opposition to each other, but each is also in opposition to those who call for the untying and expansion of aid programmes, thus making for a difficult political situation.

BRUCE DINWIDDY
This is certainly true. The British aid programme is allocated on the basis of political, commercial and developmental criteria and the government usually declines to say which criterion is followed most closely. Certainly, a government can more easily sell an aid programme to its electorate if it can show that at least some if its aid yields direct return benefits. Consequently, it will be aware of public opinion when formulating aid policies and will thus be aware of the kind of situation you describe. There is no easy answer to this, but we need to recognize, as you say, that it is a political problem.

SARAH WELLS
I would question whether it is really easier to 'sell' an aid programme to an electorate on the basis of a tied programme. The Swedish experience suggests that this is not the case. They have a larger aid programme than ourselves and seem to find it easier to convince their electorate of its value.

BRUCE DINWIDDY
I would not say that that is the case in Sweden at present. The

programme is certainly expanding, but the present increase includes a large proportion of aid tied to commodity purchase. The international development agency in Sweden is trying to persuade the government to stop this practice, but as yet has not been successful in doing this.

MILES DANBY
Is it your impression that the governments of less developed countries are now tending to view aid programmes with suspicion because of the failure of some past aid programmes and because of the obvious political as well as economic strings attached to particular programmes?

BRUCE DINWIDDY
Some developing countries have shown themselves prepared to turn down aid programmes which they do not consider advantageous, Tanzania being a notable example. But the fact is that in general, less developed countries do continue to accept whatever aid is on offer.

MILES DANBY
But wouldn't you agree that attempts to achieve a greater measure of self-reliance are increasing?

BRUCE DINWIDDY
This is true in the sense that countries are trying to diversify their sources of aid, several French-speaking African countries are examples of this trend. But there is little evidence that complete self-reliance is being pursued to any great extent. Even in the oft-quoted example of Tanzania, that country continues to accept aid from many different quarters. I think that the extent to which self-reliance is a major international trend tends to be over-estimated.

G S PURI
I think one must distinguish between aid from private sources, such as Oxfam and Christian Aid, and aid from government sources. In the latter case, quite apart from the problems of undertaking the relevant programmes, in too many instances aid is a political tool, and it has frequently been a tool in the cold war. According to an African proverb, 'when two elephants

fight, it is the grass that suffers'. In a similar way, I feel that developing countries are too often pawns in the propaganda war between east and west. In some cases there is a feeling that aid helps the donor countries more in extending their area of trade and commerce rather than the receiving countries. For example, supplying blankets to suffering people is of little value in Bangladesh in view of the climate.

BRUCE DINWIDDY
I think it is generally accepted that the aid from voluntary agencies is of better quality than most government aid, and in some cases, although small in quantity, has major effects. Whatever the disadvantage of the green revolution, the basic aim of improving food production by the breeding of high-yielding crop varieties is very valuable, and much of the effort has been financed by voluntary foundations in the United States. It isn't always true that recipients necessarily suffer from the fact that aid programmes are politically motivated. Some countries have been able to do relatively well out of playing one donor off against another!

One of the major problems in this field is that of debt-servicing: repaying past aid. Many countries could not do this without the receipt of further aid for this purpose. The fact that they need this further aid does serve to tie them to particular donor countries and it is here that they are especially vulnerable to political pressures.

JOHN TYM
I feel that we must go back to the problem of population. The present speaker mentioned that he thought it undesirable to lay too much emphasis on the population problem. It seems to me that we have spent the last 25 years working for the acceptance of the idea that curbing the growth of the world's population is essential, and that it is pointless to work towards producing more food in order that populations may continue to increase. In both of the papers so far presented, there has been a strong tendency to down-grade the importance of controlling population growth and I want to suggest that this tendency must be reversed. They are perfectly correct to say that population control programmes are ineffective unless part of comprehensive development programmes. But the reverse is also

important, namely that development programmes which do not accept the requirements of population control are also, in the long term, ineffective.

BRUCE DINWIDDY
I do not regard population control as unimportant, but, like Peter Kenyon, must stress that it must be considered as part of the whole process of world development and that the present tendency to consider it in isolation is self-defeating.

EDWIN BROOKS
What changes do you foresee in British aid policies with the advent of our membership of the European Economic Community? In particular, what is our role with regard to the European Development Fund (EDF), which, although having discernible multilateral qualities, has certainly benefited Francophone countries disproportionately in the past?

BRUCE DINWIDDY
This is all under negotiation at present, particularly in the sense that the Yaounde Convention comes up for re-negotiation later this year. The future depends very much on whether the EDF continues to be limited to these less developed countries formally associated with the EEC. I hope that this will not be the case but would not be happy about making any predictions. What I do regard as worrying is the possibility of international relationships developing increasingly on a north-south axis, with Europe concentrating on Africa, North America linked to South America, and Japan linked with South-East Asia. This would be quite different from a fully integrated commitment by the rich countries as a whole towards the poor countries.

World Trade Patterns and the Developing Countries, 1970-2000

Angus Hone

Introduction

It is best to make it clear there is very little of value in 'The Limits to Growth' and the problems of the developing countries are almost wholly ignored. Pollution is almost wholly a problem of development and the developing countries face far more serious problems of ill health, disease, food and protein shortages and persistent poverty. The Meadows' contribution has fortunately been condemned savagely by Professor Beckermann and Dr Mahbub al Haq. The subsequent debate has revealed many basic weaknesses in the 'zero growth' position and it is hoped that a clearer appreciation of the problems of developing countries has now developed.

The goal of 'zero growth' is not desirable either for the developed or developing world while there still exists considerable poverty, poor housing and a deplorable lack of opportunities for a reasonable standard of living for the poor, the old, the immigrant groups and a wide range of the industrial working class within the developed world. A radical redistribution of income within the OECD countries combined with the redistribution of wealth might solve many of the problems, but a radical redistribution of income and wealth is politically a far more formidable task than a correction of these problems under conditions of rapid growth. In the developing countries, although the UN statistics of the national income exaggerate the difference in the real living standards (see Dan Usher's work on National income differences between Thailand and the United Kingdom) [1], there are enormous differences in real standards of living and there is little likelihood of much of Africa and Asia attaining acceptable levels of per capita income within the next 100 years. There are veiled suggestions by the Meadows that the present levels of aid could be raised substantially from

$8 billion net in a 'zero growth' world, but only major cuts in defence expenditure currently running at $200 billion per year are likely to make this goal attainable.

Background

Population growth at 2.5-3.0% in the developing world has been the major factor in reducing slowly rising per capita incomes in the developing world from 1950. The growth rates of GNP between 4-5% have been satisfactory, but the bulk of the gains have been absorbed or wiped out by rising population pressure. There is every sign that until at least 2000 the population of the developing countries will grow by at least 2% per year.

Aid may grow to the 0.7% UNCTAD target for official aid but the level of private investment which has grown rapidly in the period 1967-72 cannot be expected to make a major long-run contribution to solving the developing world's heavy foreign exchange need. Financing the imports needed for industrialization and structural change in the developing world cannot rely on private investment—the costs of profits remittances and capital repayments are too high. These low expectations of aid and private investment immediately focus attention on the massive level of exports actually shipped by developing countries.

Table 1   World and Developing Countries Exports—Current Value (f.o.b. billions of US dollars)

	Developing World	The World
1960	27.3	128.0
1969	48.6	271.5
1972	65.0	354.0

(*Source: 1960-69*: UNCTAD Review of international trade and development, 1970, TD/B/309/*Rev.*, 1, p. 7. *1972*: My estimate based on price and volume growth.) 1972 was also a year of extremely high commodity prices and these levels have been exceeded in the first three months of 1973. The terms of trade paying most developing countries are no longer unfavourable, although the long-run demand prospects for a number of food, beverage and fibre crops are still very poor. The fall in the developing countries share of world trade from over 30% in 1939 to 17.8% in 1969 has not continued in 1972. The developing countries are still largely dependent on the developed market economy countries, who take almost 75% of developing country exports.

**Table 2   Developing Country Exports by Destination
(f.o.b. billions of US dollars)**

	1969	1972
Total	48.6	65.0
Developed Countries	38.0	49.5
Socialist Countries	3.0	4.5
Developing Countries	7.6	11.0

(*Source 1969*: UNCTAD Review *op. cit.*, p. 15. *1972*: My estimate.)

Rapid growth of shipments to socialist economies, which expanded from a low base around $1 billion in 1960, and expansion of intra-developing country trade have done very little to reduce the developing countries' dependence on the level of activity in the large developed market economy countries and this pattern will continue over the next 27 years until 2000.

Any cutback in the rate of raw material consumption (fuels, minerals, ferrous and non-ferrous metals) in the developed market economies will make it considerably more difficult for the developing countries to achieve the necessary import levels, which will determine rates of growth. While at the same time debt-servicing's percentage share of export earnings will rise dramatically and debt rescheduling and default will become increasingly common. The outlook for imports of food and other raw materials are not very good at the present time and would worsen if growth rates in the developed world dropped. Similarly although manufactured exports from the developing world have grown rapidly from 1960 and have continued to grow from 1966, any cutback in growth of developed countries' consumption which is growth dependent consumption, would make the programmes of 'adjustment assistance' for declining uncompetitive labour or skill-intensive industries in Western Europe, North America and Japan extremely difficult to finance; even if the present opposition of trades unions and employers can be overcome. The broad growth rates by commodity groups are shown overleaf.

It must be noted that the Generalized System of Preferences even in its limited form only applies to 18% of trade and a 50% reduction of m.f.n. tariffs on processed foods, semi-processed raw materials and metals and manufactures would have been a

Table 3　Estimated Annual Percentage Growth Rates in Value of
Developing Country Exports, 1960-72

SITC number		
0 + 1	Food and Beverages	3.5
2 + 4	Raw Materials	1.8
3	Fuels	10.0
65	Textiles	7.0
68	Non-ferrous metals	9.9
5-8 less 65, 68	Other manufactures	18.0
Total	All exports	10.0

(*Source*: My estimate from Table 7, UNCTAD, *op. cit.*, 17, and GATT Annual Reports 1970 and 1971.) The commodity composition of developing countries is still heavily biased towards non-manufactured products, although this has changed since 1960.

far more valuable concession than the limited implementation of preference systems, which in the case of many 'sensitive' products (actually, uncompetitive is a better word for these products) had already outstripped their 1972 duty-free quotes in 1969 or 1970.

Future Growth

The present structure of trade is shown in Table 4 and, using the estimated 1960-72 growth rates, the 1968 values and 1972 estimated value derived from Table 4 are shown in Table 5.

Table 4　Commodity Composition of Developing Countries Trade (%)

	1960	1968	1972
Food and beverages	29.6	25	23
Raw Materials	27.9	19.1	17
Fuels	27.9	33.6	36
Chemicals	1.1	1.6	1.8
Machinery	0.7	1.7	2.2
Non-ferrous metals	4.9	6.4	6.0
Other Manufactures	7.5	12.2	14.0
Total	100	100	100

(*Source: 1960-68*, UNCTAD, *op. cit.*, Table 8, p. 17. *1972*: My estimate. (All figures rounded.)

Table 5   Export Values by Commodity Group (in US dollar billions)
from the Developing World

	1968	1972
Food Beverages and Tobacco	11.02	14.5
Mineral Fuels	14.79	24.2
Crude Material	8.4	12.7
Non-ferrous Metals	2.81	4.0
Other Manufactures	6.84	9.6
Total	44.0	65.0

(*Source: 1968*, Table 4; *1972*, my estimates.)

The pattern of trade growth in the five years 1968-72 has been most satisfactory for the developing world. The OPEC countries have been able to negotiate substantially higher royalty payments and are now likely to own 50% of their oil fields by 1980-82. The recent boom in commodity prices has been accompanied by weaknesses in fibres, beverages and oils and fats, but there has been a steady improvement in the developing countries' terms of trade, even when oil-producing countries are excluded. A start has been made in the expansion of manufactured exports—largely without the help of the multinational corporations (G Helleiner suggests in the March 1973 issue of the *Economic Journal* that multinationals have been very important, but since Asia exports 70% of the developing countries' manufactured goods and only 15% of these exports are produced by foreign investors of multinationals there is no doubt that the role of the multinational has been exaggerated). The main concern of manufacturer exporters should be with the danger of the United States, Japan and the EEC moving towards more protectionist policies. Many cuts in tariffs will be worthless if they are rendered null and void by formal or informal government-to-government agreements, many of which are not brought to the GATT for discussion.

This paper will proceed to generate some figures for developing country exports for each of the major commodity groups for 1980, 1990 and 2000 as well as an overall growth of trade assuming that 'normal' policy conditions govern tariff and non-tariff barriers in the developed world.

Table 6 should demonstrate even to those who believe that a

Table 6 Developing Countries' Exports Assuming Differing Growth Rates
(US dollar billion in 1972 prices)

	8%	10%	12%
1972 (Actual)	65	65	65
1980	118	135	162
1990	260	340	524
2000	550	900	1,700

(*Source*: My projection based on 8, 10, 12% growth in developing countries' exports.)
(All figures rounded.)

larger transfer of *net reserves* should be made from developed economics, that politically, expansion of trade with the developing world is likely to be far less difficult and far more effective in raising developing countries' import purchasing power than aid. The value of aid at the margin will be considerable, but the present developed world GNP is around $3,000 billion at 1972 prices. This figure would increase to $10 000 billion by 2000 assuming a growth rate slightly over the 5% level. An aid programme of 1% of GNP in gross terms would amount to $100 billion per year. However, this politically ambitious target is less than 20% of the 8% of trade figure, less than 10% of the 10% growth rate and a fraction of the 12% growth. Trade is also likely to be far easier to defend politically than aid.

Subsequent tables on the sectoral growth rates will use *the 8% growth rate in developing countries' trade*. All long-term projections are hazardous and the international trade of developing countries is doubly vulnerable. First, to the international economic policies of the developed countries as they directly affect the developing world exports, but more seriously the developing countries suffer indirectly from any increase in protectionism resulting from economic bloc conflict. (Restrictions on Japanese imports to the enlarged EEC would be likely to be rapidly extended to developing country goods: EEC processed food manufacturers are already arguing that the US and Japanese processors are locating in the Philippines and Thailand to evade restrictions.) Similarly a trade war in agricultural products between the USA and the EEC will be likely to weaken the overall prospects of the developing countries' potential in agricultural investments; expansion of production capacity in fodder crops, oils and fats and meat—the major EEC import needs 1975-85—require massive long-term investments which are difficult to justify if market access and market quantity offtakes are highly uncertain. Overall, the policies on market access and the rate of growth of the developed market economies will be the key determinants of export growth in developing countries, but since the projections are in value terms, an additional danger must be pointed out. The likely growth in demand for food, beverages and industrial fibres, some metals and even a wide range of 'staple' labour-intensive manufactured items will be limited and unit prices will tend to fall if a large number of developing countries adopt the 'rational' export promotion policies embodied in the Little/Scitovsky/Scott doctrines [2]. Multinational promotion and technical support to industrial users would expand the usage of beverages (such as tea, coffee and cocoa) and fibres (such as cotton, jute and sisal), but the likely growth of most items will be limited and only carefully planned expansion of production should be undertaken.

The level of prices of manufactures will depend on a combination of 'market access', growth, and the expansion of output levels, but a concentration on a narrow range in textiles, clothing, plastic and wood manufactures etc., will produce acute market access problems and very low, if not unremunerative, prices and producers competing among themselves. These qualifications are necessary to prevent overoptimism about export prospects (the Little/Scitovsky/Scott thesis), which has replaced the dominant orthodoxy of the 1950's and 1960's (the export pessimism thesis), which was represented in the purest form by the work of Raul Prebisch at ECLA and later at UNCTAD. At the same time this paper will not only concentrate on the sectoral breakdown of the 8% assumption but on policies which developed countries could adopt to expand developing country trade.

Commodity Projections and Policies

The broad growth picture must be examined (see Table 7 below):

Table 7   1972-2000 Overall Export Trade of Developing Countries
(Total and commodity breakdowns)   (US dollars billion)

	Total	Foods, beverages and tobacco	Mineral fuels	Crude materials	Non-ferrous metals	Other manu-factures
1972 (Actual)	65	15	24	13	4	10
1980	118	20	55	17	8	18
1990	260	25	140	25	15	55
2000	550	40	340	35	25	110

(*Source: 1972* Projection Tables 5 and 6, *1980, 1990* and *2000* projection of commodity values (my projections). All figures are rounded to the nearest billion. All values in 1972 prices.)

The basic pattern of developing countries' trade confirms the growth of the pattern from 1960-70 (see Table 3). The growth of food, beverages and tobacco and crude materials is low. The growth of the fuel sector is explosive as the developed world competes for the limited energy sources. Non-ferrous metals and manufactures also register rapid growth, although it should be noted that it is in the period 1980-2000 when manufactures grow most rapidly, partially because it is somewhat optimistically assumed that sensible 'adjustment assistance' policies will be implemented in the developed world and that these will be linked with 'zero tariffs' and an abolition of quantitative restrictions. The policies will produce a considerable switch in the location of industrial activity, particularly for labour-intensive manufacturing processes.

Food, Beverages and Tobacco

There are certain items which are highly income-elastic in developed, socialist and developing economics. The most obvious growth item is *meat,* but *processed fish* such as *shrimp, tuna, crab, fresh tropical fruits and vegetables, edible nuts,* and a wide range of *exotic prepared foods* are likely to sell extremely well in the developed countries. The demand for tropical produce in the USSR and the socialist countries of Eastern Europe has grown rapidly, and with high levels of per capita income growth and a gradual relaxing of purchasing controls will grow more rapidly than in the OECD countries over the next 30 years. The potential for consumption within the developing world is immense, but family income levels are likely to remain too low to translate this into effective demand by 2000. A danger in this area is that rising domestic demand may reduce exportable surpluses. Some countries have already encountered this difficulty: Argentina in meat and India in tea and spices. Nevertheless trade in the food sector is likely to grow slowly unless the developing countries manage to expand their *meat* and *fish* production massively. In most areas of Asia and Africa this is almost impossible given the serious protein deficiency position in 1970, which is likely to worsen over the next 30 years. The major beverages: *tea, coffee* and *cocoa* are worth nearly $3 billion in 1972, but have only limited expansion prospects: the best opportunities are in the socialist countries and the developing countries, but overall growth prospects are poor. Finally, the growth of the world's tobacco industry is likely to be severely limited by the health risks attached to smoking, but in any event the United States and not the developing countries is likely to remain the dominant supplier. The projection in Table 7 assumes that the developing world will find it difficult to produce or expand its surplus of cereals, oils and fats, meat and fish in the face of acute population pressure. A gradual reduction of barriers to trade in cereals, fodder crops and sugar should act as an incentive to Latin America and East African production, but the main growth will come through rising unit values from 1990-2000 as food shortages grow. The most helpful policy instrument, which developed countries could adopt would be the abolition of tariffs on processed and packed foods and beverages from the tropical world when in competition with temperate produce. A large part of these products will continue to be processed in the importing countries because it is the marketing costs and skills, and not the production capacity, which affects the location of processing facilities. The enlarged EEC has an opportunity to show that it is not just a self-help society for the rich when the original EEC and the United Kingdom's Generalized System of Preferences are merged on 1st January 1974 by offering at least a 50% cut in the present high duties on processed and packaged foods. Professor Hans Singer has rightly emphasized that the net value-added, the net foreign exchange earnings and the employment value from $1,000 of processed food imported from the developing world is many times greater than that associated with a similar value of imports from clothing, electronics or similar industries. Developed countries anxious to make good their often-expressed statements on employment problems of developing countries would do well to begin cutting their duties or levies on processed and packaged food products.

Mineral Fuels

Nineteen seventy-two was the year in which the world discovered that it was likely to suffer an energy crisis in the 1980s. This crisis was brought home to the Western oil companies in 1971/72 through the doubling of oil prices and negotiations between the producer nations and the oil companies leading to a long-term production sharing agreement. The growth in the mineral fuel sector from $24 billion in 1972 to $140 billion in 1990 and $340 billion in 2000 partly reflects the 8-10% growth in demand and at the same time a 10-12% likely growth in oil prices. Oil reserves are inequitably distributed and the export earnings of the oil producers are unlikely to be given as aid to their less fortunate developing countries except on a rather narrow and limited base (the Kuwait Development Fund) or on political grounds (Libyan assistance). However, these measures may improve from 1980 when the full extent of the earnings flow is realized. Over the period 1972-80 the largest financial flows from oil countries to non-oil countries are likely to be through World Bank loan subscriptions.

The interaction between growing demand and shortage and high prices is just the kind of feedback which is excluded from 'The Limits to Growth'. First, the higher prices will stimulate an intensive search for new energy sources (oil and non-oil), which may eventually damp the rate of price increases and the high prices will eventually produce a long-term reduction in energy usage. The absence of the feedback mechanism of rising prices is severely criticized by Mahbub al Haq in his recent excellent article [3].

The most serious consequences of rising oil prices will be on developing countries with no fuel reserves. These countries' industrialization drives will be stunted if they are forced to pay the full price for mineral fuels. The OPEC countries would do well to consider a 'dual price' system for oil: the developed countries would pay the full international price and the developing countries a considerably reduced price provided the entire quantity was consumed domestically. Any cutback in developed country growth will immediately affect oil incomes in the developing world and Table 8 gives some idea of the resource transfer required to substitute for oil earnings:

Table 8   Oil Earnings in US dollar billions

1962	24
1980	55
1990	140
2000	340

(*Source*: Table 7.) The difficulty of organizing transfers of aid on this scale will surely be realized.

Crude Materials

The growth of crude materials (fibres, timber, leather, minerals, etc.) are critically affected by the development of synthetics. The major fibres (cotton, wool, jute, sisal, etc.) have all been threatened by plastics, now almost wholly oil-based in production. The short-term prospects for jute, wool and sisal are poor, but cotton has made a considerable comeback. The longer-term prospects for fibres will depend on the increase in oil prices, which will certainly affect polymer costs over the next 20 years, and on the production costs of natural fibres. Our projection of earnings are based on the likelihood of higher prices for all natural products as world incomes rise and a shortage of natural materials develops. The gains in crude material exports of developing countries will affect almost all developing countries as fibres, leather, timber and minerals are widely produced in Africa and Asia and even in some of the poorest countries. Again, the gains will be greatest for the developing countries if world growth continues at a high rate, but the developing countries must pay attention to the needs of raw material conservation, reafforestation and the competition between fibre and food crops. Many developing countries have made extremely unfavourable agreements with international mining groups (which they should renegotiate), or have been selling their timber reserves too cheaply or exploiting these reserves indiscriminately without any provision for replanting.

Non-ferrous Metals

The prospects for growth in non-ferrous metals (see Table 7) are not very striking. There is great danger of substitution within the metals themselves and with plastics. Copper and aluminium are directly competitive in many electrical uses. Tin and zinc are being attacked in their traditional markets by new coating materials while aluminium's growth has been cut by plastic substitutes. The key question for most developing countries is the net export earnings or *net retained value,* which they keep from a unit of mineral exports. The OPEC countries have succeeded in renegotiating the various treaties in oil, but the world production of non-ferrous metals is much more equally balanced between the developed and the developing world and the growth in demand has been much slower. Prices will rise, but new mining and recovery techniques will probably make it possible for developed countries to exploit thinner ore bodies if there are too many attempts to follow OPEC's example.

Other Manufactures

This area has been chosen by the new 'export optimists' and it is likely to grow very rapidly *if* economic rationality prevails. At present the EEC is importing immigrant labour to manufacture labour-intensive goods. The social consequences of this policy are gradually beginning to be realized. First, there are very serious problems of conflict, which arise from competition between EEC citizens and immigrants for housing, social services and educational facilities. Second, the backward regions within the EEC develop only slowly as the labour-short industries at the centre are not forced to move to peripheral regions. Finally, the problem of unemployment in the immigrants' countries is only partially relieved as many of the most educated, trained and ambitious workers leave. The future policy of the EEC should follow more closely those of Japan and the United States who have expanded their production facilities or their buying organizations throughout Asia and to a limited extent in Latin America. The policy of 'zero tariffs' in manufactures will produce many temporary problems of unemployment in the developed world in the following major industries:

> Textiles
> Clothing
> Processed Food
> Wood Manufactures
> Electronics
> Plastics
> Footwear and Leathergoods
> Electrical Engineering
> Automotive Parts

These industries already face considerable pressure and are at present protected by one of the following measures:

> High Tariffs
> Quotas
> Health and Sanitary Regulations
> Voluntary Restraints

All these measures restrict competition and remain politically viable only because there is no carefully worked out 'adjustment assistance' package to protect workers. It is futile for the political economists to protest that there is no need for assistance because the developed country has already benefited through lower prices for its imports. An EEC package should include the following provisions:

1. 2 years at present level of take-home pay, *followed by*
2. 5 years at 75% of present level;
3. 3 years retraining or education at the country's expense;
4. 100% payment of all removal expenses;
5. Guaranteed housing in new occupations/new location.

It will be argued that such a package is too generous and too costly, but the present social and economic costs of immigrant workers combined with the continued operation of inefficient and high cost industry is several times the cost of a generous 'adjustment assistance' package. The growth rates shown for manufactured exports can easily be attained by the developing countries without the aid of multinational corporations and this development of a national pattern of location would do a great deal to alleviate the employment problem of the urban areas in many developing countries.

Export Prospects

There is no reason to be pessimistic about the export prospects of developing countries provided the developed world continues to grow, avoids thermo-nuclear war and pursues quasi-rational economic policies. Slow growth in the developed world would intensify protectionism pressures and make it extremely difficult for exports to grow throughout the developing world. The supporters of 'zero growth' should examine Table 7 closely and consider whether there is any political prospect at all of making aid available on the required scale to permit 5% growth (the minimum target) in the developing world.

Developing Country Imports

These export levels are required to sustain imports of intermediate and capital goods, which are vital for growth. Official aid in 1972 at $7 billion is running at approximately 10% of developing exports, but there are considerable amounts of debt repayment, which must be made before new imports can be considered. Most developing countries probably will continue to be short of foreign exchange throughout the period and any increase in aid will be welcome, but, as Table 9 shows, the levels of aid are unlikely to maintain a 10% ratio with exports.

Developed Country Policies

There are a number of policies, which the developed world can adopt to improve the trade prospects of the developing world:

(1) *A removal of protective quotas or tariffs on agricultural production*
    The loss to the developing countries was estimated by

Table 9 Implied Aid Levels at 10% of Exports (Official Aid)
(US dollars billions)

	Exports	Aid
1972 (Actual)	65	7
1980	118	12
1990	260	26
2000	550	55

(*Source*: Table 7.) These levels of aid are widely regarded as unattainable in the present political atmosphere and so the percentage of imports financed by exports will almost certainly have to rise sharply.

the FAO in their 1970-80 Commodity Perspective at 6% of national income. The EEC and Japan are the worst offenders and a removal of quotas on *meat, fodder crops, rice, sugar, fruits and vegetables* and substantial cuts in tariffs on all processed agricultural products would be a major incentive.

(2) *'Zero tariff' for manufactured exports*

The GATT talks in the autumn of 1973 will attempt to work out a programme for moving towards zero tariffs, which will erode the advantages (limited though they are) of the Generalized System of Preferences. The best offer would be a revised GSP offering developing countries zero tariffs on all commodities from 1975 without any quota limitations.

(3) *Abolition of formal or informal quota agreements*

The zero tariff regime would mean very little if quotas could not be abolished over a definite time period (say 3-5 years). Clearly all quotas should be first replaced by an 'equivalent tariff', which can then be cut each year.

(4) *'Adjustment assistance' policies* (see Other Manufactures).

All the suggested measures will fail if there is no carefully worked out package of policies designed to assist workers and industrialists, who suffer severely from developing world competition. The policies were explored in detail, *above,* although limited compensation and generous bank loans would have to be made available to industrialists to allow them to establish new businesses—the rate could be on a sliding scale which would pay substantially more if the new industries were located in backward regions of the EEC, United States or Japan.

These policies relate to trade, but one final area must be covered. The Special Drawing Rights or SDRs at present provide only $1 billion to the developing countries, but they could be expanded and linked to purchases of developed country goods. By 1980 SDRs at $3 billion would be a very valuable form of aid, which would directly expand the developing world's purchasing power. However, they are unlikely to expand beyond that level because of fears of adding to world inflation.

Conclusion

The importance of trade in world development exceeds that of aid. A slower rate of world growth would directly reduce the importing capacity of the developing world without doing anything to ensure aid flows at a compensatory level. The opponents of growth would do better to spend their time examining the mechanisms for the transfer of resources from 'rich' to 'poor' countries and their political viability. There are trade policies which would help far more dramatically than a 1% level of official aid, which appears still to be unattainable.

References

1. Dan Usher, *The Price Mechanism and the Meaning of National Income Statistics*, Oxford 1968.
2. Ian Little, Tibor Scitovsky and Maurice Scott, *Industry and Trade in Some Developing Countries: A Comparative Study*, Oxford 1970.
3. Mahbub al Haq, 'Limits to Growth: A Critique', *Finance and Development*, 4, December 1972, pp 2-8.

Postscript

It is true that there is a clear trade-off between increased flows of aid and reduced growth. The developing countries are not tied to 8% growth of world trade as a means of financing their development efforts. Table 10 below shows the possibility of different patterns of growth:

Table 10   Total Resource Flows to Developing Countries

| | Base 1972 $65 Billion | | | | Base $7 Billion | | |
	2½%	4%	Exports 8%	AID 1%	2%	3%	4%
1980	80	90	118	12	24	36	48
1990	105	120	260	26	52	78	104
2000	130	200	550	55	110	165	220

This comparison of trade growth at 2½%, 4% and 8% shows the sensitivity of trade growth to cuts in overall growth. It must also be pointed out that aid at 4% of gross national products in developed countries does not fully compensate for losses in trade. It does, however, give a good idea of trade-off between trade growth and aid growth required if developed countries decided to adopt lower growth options.

Discussion

JOHN TYM

Mr Hone mentioned 'growth' and 'zero growth'. It seems to me to be important that we have exact definitions of these two things and I would appreciate his giving such definitions.

ANGUS HONE

I accept that what is meant by growth is a subject of dispute among development economists. I would have thought that in the context of a developing country if you have an increase in consumption per head and if that consumption is more equally distributed than previously then that is what is meant by growth, and in such a context the growth represents an improvement in the state of development of that country. As Peter Kenyon and others have pointed out, much economic growth in less developed countries has resulted in less equal distribution of consumption, and the poor have often not benefited at all.

Zero growth for the developed world has a rather different context and implies no increase in consumption per head, as valued at present market prices. That doesn't mean that you don't change the distribution within the country concerned. You can have increased expenditure of money per capita and still have zero growth in terms of consumption of material resources. For example, more money might be spent on parks and less on consumer goods.

If I could add one last point, I would like to see growth measured in terms of the items consumed by the population of the country concerned. For example, in India in terms of rice, millets and the like and in Britain in terms of cars, holidays, etc. So far as less developed countries are concerned, growth must be measured in terms of what the people need to be able to live in at least tolerable comfort. Zero growth may be desirable in developed countries, based on different growth indicators when compared with less developed countries where it is very definitely arguable that it is not desirable.

G S PURI
Towards the end of his paper the speaker mentioned pollution—are we going to export pollution to the developing countries? I would like to raise a second point, about the role of Japan as a major trading nation in the less-developed world, and the possible future role of this traditionally large and powerful trading nation. One further point is that we have heard a great deal about poverty, but nobody seems to have raised the point that India for instance, would be a much more buoyant country economically if all the gold and jewels owned by Indian families could be mobilized. Furthermore, our speaker has not mentioned the trading in gold which is an important factor.

ANGUS HONE
I think on pollution that there is a definite danger that pollution will be exported. I think this is an inevitable part of the process of development. It is very difficult to run a slaughter house without polluting water—it is not impossible but I think it is difficult—so I think developing countries will have to watch for the dangers of pollution in the manufacturing processes.

On Japan, I don't exaggerate the dangers of their relatively high level of wage costs. Already a number of industries from Japan are locating in countries outside Japan—in Taiwan, Korea, Indonesia and Hong Kong and India. Japan at the moment is buying large quantities of hand tools from India—they are imported with a Japanese brand name on them and then exported to the United States.

On the question of reserves of gold and diamonds, a recent report in India indicates that the illegal smuggling of gold, diamonds and synthetic rayon etc., was running at a level, plus over- and under-invoicing, of 160 million dollars a year. The value of India's gold reserves almost certainly exceeds 10-12 billion dollars; the difficulty is getting access to it. During the China war, a gold restoration scheme was not very successful—it is estimated that it got less than 0.05 of 1% of the total gold that was available in India. The gold of course is held as a precautionary measure, and gold is the best precautionary measure you can have.

The Global Ecosystem

Palmer Newbould

Introduction

On spaceship Earth there is an interrelationship between population, resources and environment. Population is a demand factor, making demands upon resources and environment. The demand made is the product of numbers and expectations, both of which are growing rapidly at the present time. The end product I take to be human welfare, which probably involves satisfying most but not all of the expectation. Resources are often classified into renewable, like soil and plants, or non-renewable, like fossil fuels or metals. Both sorts, however, have fairly finite limits to their exploitations. The environment may be regarded as having a finite if somewhat elastic carrying capacity for people and their activities, including their wastes and pollution. Beyond this limit, serious problems, such as the breakdown of life-support systems, may occur. There is also something perceived by the population as a good environment and therefore as a component of the quality of life and hence of human welfare. Certain human activity tends to cause a progressive decline in the quality of the environment and this occurs well before the total ecological breakdown mentioned above.

In general the harder population pushes on resources the more severe will be the environmental consequences. This is because more energy (or money) is expended to achieve a particular level of resource availability and much of this energy is involved in environmental change and often pollution. The mining of metals in Snowdonia and the winning of oil from the North Sea or Alaska are expressions of the scarcity of these commodities. Intensified food production has greater environmental consequences than more extensive systems. Economic development in the context of this conference presumably involves the allocation of a greater share of resources and of use of the environment to the people of a developing country, so that more of their expectations are realized. It sometimes has, coincidentally, the unfortunate effect of increasing population growth, at least in its earlier stages. It is based on capital investment which in ecological terms means diverting some of the resources available for use now towards increasing the availability of resources at some future time, e.g. using energy not to generate more electricity from existing power stations but to build more power stations; alternatively putting up the price of electricity now and using the profit to build more power stations.

There is in fact here a close parallel with the development of ecosystems as described by Odum (1969). Energy fixed by photosynthesis is partitioned between the maintenance of all the living organisms in the ecosystem and the accumulation of organic matter, living and dead. Where man takes a crop, as for example in a grassland, this tends to reduce the accumulation of organic matter. The grassland remains grassland all the time it is cropped, but when cropping ceases it accumulates more organic matter and becomes forest. The final stage, the stable climax ecosystem is where the entire energy input goes in maintenance.

If we now compare natural and man-managed systems, the man-managed systems mainly have a dual energy input, part directly from solar radiation and part indirectly from fossil fuels. The extent to which we can continue the fuel energy input depends upon the future development of nuclear energy, especially nuclear fusion energy. If we assumed a constant level of fuel energy input, then perhaps the partitioning between accumulation of artifacts and the maintenance of existing systems must move, over the next hundred years, towards the latter and the stable economy. This implies the philosophy of restricting demand as well as increasing supply.

All this is very generalized, in fact the sort of generalization that, if unsubstantiated, gets ecology a bad name. I propose, therefore, to try and look more closely at some of the ecological constraints upon economic development. In general, development is likely to involve major changes in the pattern of use and management of land and other resources, and an increase in the amount of pollutants released into the environment. The problem is to determine what changes and what amounts of increase in pollutants are acceptable, both in particular instances and on a global scale. The human welfare accruing from development has to be evaluated against any losses of human welfare resulting from changes in land use and in the physical environment. There are no absolutes, nor is the prediction of ecological consequences at all precise as yet. But where one is dealing with life-support systems and with irreversible change, caution is needed. I will discuss eight aspects of the interaction of development and the biosphere.

Agriculture and Forestry

Development will bring a need for an increased, sustainable yield of food and fibre, both for the developing population and for export. Food means essentially protein as well as carbohydrate. Food yield is a product of area x intensity. Limits to both area and intensity are set by climate, terrain and soil fertility and further limits to intensity are set by economic factors. Intensification means increasing the input of fuel and/or human energy per area of land while the solar energy input remains the same. Thus in the USA the fuel energy consumed directly and indirectly by agriculture is in energy terms about five times the food produced, five calories of fuel for one calorie of food (Perelman, 1972). Where the factor limiting productivity is difficult to modify, e.g. solar radiation or CO_2 concentration of the air, the law of diminishing returns operates. Also the most suitable areas for intensification are already so used. Intensive agricultural systems are inherently unstable (their stability derives from their human inputs) and create many practical problems such as increased pests and diseases, side effects of pesticides and antibiotics, the development by pests of immunity to chemical pesticides, side effects of heavy applications of fertilizer, loss of soil organic matter and structure, problems of slurry disposal, soil compaction by heavy machinery and so on (Alexander, 1973). These environmental consequences may therefore set limits on the intensification of agriculture. The limit to the amount of inorganic nitrogenous fertilizer used may be set by the proportion which reaches rivers and lakes.

Of course human survival is utterly dependent on increasing intensification of agriculture. But there are limits to this. There

is some potential for increasing protein yields from inherently
unproductive land by improved systems of animal husbandry
including domestication of wild animals, game cropping, the use
of greater mixtures of animal species and perhaps some use of
fertilizers and herbicides. In terms of world food needs the
benefits will be small. The same amount of extra meat may
derive from doubling the yield of a hill farm or a game ranch as
from a 5% increase in production from a fertile lowland farm.
The latter will almost certainly be easier to achieve. More
drastic attempts to upgrade the productivity of infertile sites
may well be disastrous resulting in soil impoverishment and
erosion (Dasmann et al, 1973). It also seems likely that man will
continue to use cellulose, if not wood as such, as a raw material.
Man-made fibres and plastics mainly derive from oil and coal
products, which will run out, oil in perhaps 50 years and coal in
200 to 300 years. Therefore considerable land areas will
necessarily be allocated to forestry or some other cellulose-
producing system. Continuously cropped cellulose systems will
require some replacement of fertilizer and in mountainous or
high rainfall areas the cropping programme must not leave the
soil bare. Clear felling may be unacceptable. So development
may require more food and fibre production. Extension of
productive area and intensification may both have environmental consequences.

Pollution

The biosphere must continue to disperse, absorb or metabolize
pollutants released into it so that they do not accumulate
anywhere in harmful quantities. For example, rivers are used to
oxidize organic effluents. Overloading creates so great an
oxygen deficiency that the system can no longer carry out its
oxidation function efficiently as well as becoming useless for
several other functions like water supply and angling. The
atmosphere disperses and the soil eventually absorbs and
metabolizes the sulphur dioxide released from chimneys.
Increasing SO_2 production may cause soil and water to become
progressively more acid. In Sweden (Anon, 1972) a reduction of
forest growth of 0.3% per year from 1950-65 is attributed to
this cause and there is concern lest some lakes become too acid
to support fish populations.

The oceans are the great sink for all pollutants such as chlorinated hydrocarbons, PCBs (Jensen, 1972) and mercury (Harriss, 1971). Accumulation of solids at the bottom of deep oceans is probably harmless. On the continental shelf it may be beneficial or harmful to inshore fisheries according to the proportion of fish food and fish poison involved. Marine pollution is already a serious threat in small landlocked seas such as the Baltic or the Black Sea. The Mediterranean has a large population on its shores. The large oceans are less threatened in general though Thor Heyerdhal suggests that surface pollution of the Atlantic has increased very greatly over the past 20 years. One serious threat is the possible build-up in the surface waters of a substance which might significantly reduce the levels of photosynthesis of the phytoplankton. Possible candidates are DDT, PCBs or some derivative of petroleum. Assessments of this threat vary greatly (Lawton and McNeill, 1972; George, 1971; Blumer *et al*, 1971). The solubility of DDT in sea water (c 1.2 ppb) is below the level at which any effect on the photosynthesis of phytoplankton has been demonstrated. However, phytoplankton cells certainly accumulate DDT. The DDT content of phytoplankton off the American coast has trebled between 1955 and 1969 despite the declining use of DDT on the adjoining land area (Cox, 1970). If photosynthesis were reduced, this would affect not only the marine ecosystem itself but also the absorption of carbon dioxide and production of oxygen, thus affecting the composition of the atmosphere.

Radio-activity, potentially the most serious pollutant of all, is probably the best controlled at present (Scott Russell, 1969). Providing the world avoids the use of nuclear weapons the other radioactive hazard is that of accidents deriving from power stations based on nuclear fission.

Hydrology

It is necessary to retain an equable hydrological regime in which there is reasonably high water storage potential in the catchment area reducing the prevalence of damaging floods and equally of droughts. Such a hydrological regime should also reduce the prevalence of severe soil erosion, thus maintaining soil fertility and reducing silting downstream, especially as it affects reservoirs. In practice this means, as a minimum, maintaining vegetation cover on catchment areas; in many cases some of this vegetation should be forest. The restriction on clear felling mentioned above also comes in here. But the effect of catchment management on hydrological regimes remains a highly controversial subject (Penman, 1963; Dasmann *et al*, 1973; chapter 7).

Wildlife

It is necessary to retain a reasonably high diversity of plants, animals and landscape components for their amenity value. While this is difficult to quantify it is increasingly regarded as a component of the quality of life, and not only by middle class protection societies. This diversity will help by its intrinsic interest and its physiognomic diversity to absorb some of the rapidly growing requirements for outdoor recreation. Note that it is diversity, not individual species, that must be conserved. One cannot halt evolution at a particular point in time.

The threat to wildlife comes mainly from habitat destruction by, for example, forest clearance, land drainage or building works. Improved access and an increasing human population with firearms may allow direct persecution of some species. The use of chlorinated hydrocarbon pesticides may cause serious reduction in some predator populations (Wurster, 1969). Some people would also hold that there is an ethical case for trying to ensure the continuing existence of the same diversity of plant and animal species as now, since they have equal entitlement with man to survival. This may be contrary to the teaching of Genesis, chapter 1.

Overheating

It has been claimed that the ultimate limit to the earth's carrying capacity will be set by disruption of the heat balance leading to overheating. This must presumably involve the inability of the earth to dissipate all the extra heat when man's energy requirements are mainly met by nuclear energy. It will not happen while we are still dependent on fossil fuel energy because the resource is not great enough. I believe this particular hazard to be very remote. Using Deevey's (1971) unit the geocalorie (5.2×10^{18}) (one calorie per square centimetre of earth's surface) solar radiation of the order of 177×10^3 geocalories per year results in photosynthetic production of 177 geocalories supporting livestock respiration of 1.17 geocalories and human respiration of 0.58 geocalories. By 1980 this human activity will be supported by thermal energy production from fossil fuels of the order of 14.8 geocalories. The thermal energy produced is therefore about 0.01% of the solar radiation and unlikely to affect the earth's heat balance greatly. Even a 10-fold increase would have no significant effect.

Climatic Change

As an extension of possible overheating it is clearly undesirable that man should cause major climatic changes which are contrary to his own interests. Two that have been suggested are the warming up of the earth due to the increasing CO_2 concentration which results from combustion of fossil fuels and perhaps also from forest clearance (Johnson, 1970) and the cooling off of the earth due to the increased dustiness of the earth's atmosphere resulting from a variety of human activities (Wendland and Bryson, 1970). It is perhaps fortunate that these two changes operate in opposite directions. There is also the danger that the amount of ozone in the upper atmosphere might be reduced by the water vapour and oxides of nitrogen in the vapour trails of supersonic aircraft (Johnson, 1972; Crutzen, 1972). This would allow more ultraviolet radiation to penetrate which would have destructive effects.

In none of the three instances mentioned are the predictive equations especially accurate. The Royal Commission on Environmental Pollution (1971) felt that the increase of CO_2 concentration might produce an average rise in temperature of 0.1-$0.2°C$ by the year 2000. A 10% increase in atmospheric turbidity, which is about the highest considered likely might cause a general climatic cooling of the order of $1°C$. Both these figures are within the sort of background climatic fluctuation which might occur anyway. Similar conclusions emerge from the MIT study (MIT, 1970).

Water Supply

Arising from the discussion of pollution and hydrology above, man requires a continuing supply of good quality water.

Development is likely to increase the water use per head of population and to create agricultural and industrial water demands. The limits to the treatment of polluted water are part economic, part technological. At present, for example, the de-salination of sea water can only be economically justified in special circumstances. Development is likely to increase water pollution, decrease natural water storage and increase the demand for clean water. Any development plan must resolve this dilemma. It is clearly a dilemma that the USA has by no means solved (Anon, 1973).

Flexibility

It is desirable to keep as many land and resource use options open and reversible for succeeding generations as possible. Many decisions are currently made on inadequate data and it is a feature of the present age how many major decisions appear, a few years later, to be wrong, e.g. Concorde Maplin, the run-down of the railways, the urban motorways, high-rise flats, the ground nuts scheme, the whaling industry, etc. Some land or resource management policies are more easily reversible than others and clearly, other things being equal, reversibility is highly desirable. Dasmann et al (1973) distinguish six sorts of land-use option in a tract of land not yet open to human use.

1. It can be left in a completely natural state and reserved for scientific study, educational use, watershed protection and for its contribution to landscape stability;
2. It can be developed as a national park or equivalent reserve, with the natural scene remaining largely undisturbed to serve as a setting for outdoor recreation and the attraction of tourism;
3. It can be used for limited harvest of its wild vegetation or animal life, but maintained for the most part in a wild state—serving to maintain landscape stability, support certain kinds of scientific or educational uses, provide for some recreation and tourism, and yield certain commodities from its wild populations;
4. It can be used for more intensive harvest of its wild products as in forest production, pasture production for domestic livestock, or intensive wildlife production. In this case its value as a 'wild' area for scientific study diminishes but it gains usefulness for other kinds of

scientific and educational uses. Its value for tourism and outdoor recreation diminishes but is not necessarily lost; its role in landscape and watershed stability is changed, but may be maintained at a high level.

5. The wild vegetation and animal life having been removed in part, it can be intensively utilized for the cultivation of planted tree crops, pastures or farming crops; or

6. The wild vegetation and animal life having been almost completely removed, it can be used for intensively urban, industrial or transportation purposes.

They point out that choices 1-3 leave open the option to change from that use to any of the other five uses. Choice 4 makes it difficult but not impossible to restore the land to choices 1-3—while choices 5 or 6 largely prohibit any change in option.

Two principles might be enunciated here. The closer the management policy is to the natural situation, the more options are left open. The more fertile and favourable the environment, the greater is the resilience of the vegetation and the wider is the range of available management options. In extreme environments, the arctic, the desert or the mountain top, the vegetation has little resilience and there are few, if any, options and those there are do not depart far from the natural condition of that area. The nutrient impoverishment of many tropical rain forest areas reduces their management options considerably.

Conclusions

In summary, therefore, economic development may involve:

the need for increased food and fibre production and the environmental consequences of this; increased generation of domestic, agricultural and industrial pollution and reduced environmental containment of this; forest clearance changing the hydrological regime; a reduction in the area occupied by more natural ecosystems, in the diversity of fauna, flora and landscape and in the potential for general amenity and outdoor recreation; possible modifications of climate; an increased demand and reduced potential supply of clean water; a reduction in the land and resource use options open to future generations.

None of these consequences are over-riding but all have to be taken into account in any proposed plan for development. All are interrelated and this, above all, is the ecologist's message. Man and his environment are a series of interrelated systems, physical, ecological, social, economic and political. A change in one system affects all the others. But the interrelationships between the systems and the functioning of particular systems are both so complex that prediction is as yet very imperfect. Some of the effects listed are global in the sense that the circulation of the atmosphere and the oceans are global and trade in food and other resources is international. The ecological consequences of development have therefore to be considered at two levels, the specific local impact of a particular scheme and the overall global impact of all the development schemes added together. Certain zones of the world are particularly susceptible to damage. Pressures are especially high in coastal areas (Ketchum, 1972) and the humid tropics are often (and wrongly) seen as an area of great untapped potential.

My own belief is that however successful population policies are, the world population is likely to treble before it reaches stability. If the expectations of this increased population were for example, to emulate the present life style and resource use of the USA, the demand on world resources would be increased approximately 15-fold; pollution and other forms of environmental degradation might increase similarly and global ecological carrying capacity would then be seriously exceeded. There are therefore global constraints upon development set by resources and environment and these will require a reduction in the per caput resource use and environmental abuse of the developed nations to accompany the increased resource use of the developing nations, a levelling down as well as up. This conflict cannot be avoided. It became the central theme of the United Nations Stockholm Conference on the Human Environment (see for example Aaronson, 1972).

References

Most of the topics briefly referred to in this paper, and many others are fully discussed in:

FARVAR, M T and MILTON, J P (1972). *The careless technology: ecology and international development.* New York: Natural History Press, Doubleday & Co.

and also, at less length, in:

DASMANN, R F MILTON, J P and FREEMAN P H (1973). *Ecological principles for economic development.* London: Wiley.

Other references referred to in the text are:

AARONSON, T (1972). World priorities. *Environment 14* (6), 4-13.

ALEXANDER, M (1973). Environmental consequences of increasing food production. *Biol. Cons.* 5, 15-19.

ANON (1972). Sulphur pollution across national boundaries. *Ambio 1*, 15-20.

ANON (1973). The wastewater tide ebbs slowly. *Environment 15* (1), 34-42.

BLUMER, M *et al* (1971). A small oil spill. *Environment 13* (2), 2-12.

COX, J L (1970). DDT residues in marine phytoplankton: increase from 1955 to 1969. *Science 170,* 71-3.

CRUTZEN, P (1972). SSTs—a threat to the earth's ozone shield. *Ambio 1,* 41-51.

DEEVEY, E C (1971). The chemistry of wealth. *Bull. Ecol. Soc. Amer. 52* (4), 3-8.

GEORGE, J D (1971). Can the seas survive. *Ecologist 1* (9), 4-9.

HARRISS, R C (1971). Ecological implications of mercury pollution in aquatic systems. *Biol. Cons.* 3, 279-83.

JENSEN, S (1972). The PCB Story. *Ambio 1,* 123-31.

JOHNSON, F S (1970). The balance of atmospheric oxygen and carbon dioxide. *Biol. Cons.* 2, 83-9.

JOHNSON, F S (1972). Ozone and SSTs. *Biol. Cons.* 4, 220.

KETCHUM, B H, ed. (1972). *The waters' edge: critical problems of the coastal zone.* MIT Press, Cambridge, Mass. and London.

LAWTON, J H and McNEILL, S (1972). Pollution and world primary production. *Biol. Cons.* 4, 329-34

Mass. Inst. Technology (1970). *Man's impact on the global environment.* MIT Press, Cambridge, Mass. and London.

ODUM, E P (1969). The strategy of ecosystem development. *Science* 164, 262-70.

PENMAN, H L (1963). *Vegetation and Hydrology.* Commonwealth Bur. Soils Tech. Comm. 53.

PERELMAN, M J (1972). Farming with petroleum. *Environment 14* (8), 8-13.

Royal Commission on Environmental Pollution (1971). *First Report.* HMSO.

SCOTT RUSSELL, R. (1969). Contamination of the biosphere with radioactivity. *Biol. Cons.* 2, 2-9.

WENDLAND, W M and BRYSON, R. A. (1970). Atmospheric dustiness, man and climatic change. *Biol. Cons.* 2, 125-8.

WURSTER, C F (1969). Chlorinated hydrocarbon insecticides and the world ecosystem. *Biol. Cons.* 1, 123-9.

Discussion

J GRANT
With regard to the relationship between vegetation and climate, it may be true to say that a unit of vegetation on earth corresponds to a unit of ice at the Pole, so that if the vegetational condition of the earth is reduced the temperature will rise, and if it is increased the temperature will go down, as shown in recent glaciations.

P J NEWBOULD
I don't think the relationship is all that close; of course cities are creating heat islands, but these are very local. The pluvial period in regions near the equator seems to correspond with the glacial period in regions further from the equator. In Africa, there is a system of pluvials and interpluvials corresponding with glacials and interglacials. At the moment I don't think that we are causing major climatic changes at a significant rate.

D HALL
Supposing there is a limited amount of money which the Overseas Development Administration can give to a particular country, and the choice of use was one between industrialization on the one hand, and a study of the ecosystem in depth on the other, could we have some discussion as to which of these two would get priority.

P J NEWBOULD
What I am saying is that plans for development should have environmental safeguards built into them as far as possible, so I don't see life in terms of that single choice. The World Bank for instance examines much more carefully the environmental consequences of a scheme before lending money, and to that extent there is a shift of money towards ecological study.

A R VANN
You mentioned that before population stabilization occurred, there might be a 15-fold demand on environmental resources. Are you then saying that the peoples of underdeveloped nations aspire to our level of consumption?

P J NEWBOULD
With the present level of communication, an impression is being created of what is the 'good life', and if so many people get this impression, that is the point at which I start to get worried. The population cannot, I submit, stabilize at any level short of 10,000 million other than catastrophically. So assuming this level of population, we can say that it is expectations or demand which may have to be manipulated.

J TYM
Mr Hone stated that pollution is likely to be the next fad. I submit that, with regard to pollution of fresh water systems, the point is not simply to provide people with a little cleaner water than they have had so far, as Mr Hone suggested, but that we might be able to derive a little more protein from water resources which are now closed to us as a result of the kind of growth some people have tended to support for the past 50 years. As a result we have had to import protein from the developing nations for our mounting populations.

P J NEWBOULD
I would agree to that. In Northern Ireland one of our growth industries is trout farming, and it is fairly efficient in terms of protein conversion. However, there is a lot of inefficiency in the production of high-protein food which you give to the trout. This farming is, of course, dependent on large supplies of clean water, and is a good example of protein production totally reliant on clean water. There are probably even better examples in developing countries.

A HONE
I am afraid that it is a choice between clean water and trout in developed countries and no water at all for people in developing countries. We are talking about the allocation of resources, and unless there is a very big change in our ideas and assumptions

about development, the resources which are likely to be spent on environmental control are likely to detract from the resources which are given to overseas aid or trade. They may of course merely prevent the levels of resources devoted to aid and so on from rising.

J TYM
That is not relevant!

Ecological Effects of Current Development Processes in Less Developed Countries

Kenneth A. Dahlberg

Introduction

It would be intriguing to take a position rather like that of Gandhi and argue that it is the rich Western countries that are really the 'less developed' ones as they have not been able to come to grips with their masses of material goods either spiritually or ecologically. The temptation is the greater because many ecologists are doing much the same thing: they are suggesting that we, the rich Western countries, must change our basic attitudes towards nature, our modes of production, many of our institutions, and many of our comfortable habits. The challenges which these 'subversive scientists'[1] present to our traditional ways are strikingly similar to the challenges which conventional economists have presented to the non-Western world in urging them to 'develop'. It is only when we think about the degree and difficulty of change that is being asked for by the ecologists (as well as the social and institutional resistance to such changes) that we can begin to appreciate the ways in which conventional development processes challenge the poorer countries of the world.[2] Such a line of argument would also more clearly stress that it is the rich countries that are the greater sinners in terms of pursuing anti-ecological policies and technologies and that, given the wide-ranging impact of the rich upon the poor, it is really in the rich countries that the major changes must come if we are going to learn to live within the limits of our planet.

[1] P Shepard and D McKinley (eds.), *The Subversive Science: Essays Towards an Ecology of Man*, Boston: Houghton-Mifflin, 1969.

[2] The parallels are not complete, for the ecologists represent an *internal* challenge, rather than one with external origins. Also, the goals are rather different: the economists suggesting quantitative goals relating to material progress, the ecologists stressing qualitative goals—which however must be consistent with, as well as encourage, survival and adaptation.

While the above 'switch' in definitions will not be pursued, it will be argued that conventional understandings of 'development' must be modified to include the recognition that what is really involved is a process of transferring or, more precisely, imposing Western values, concepts, technologies, and institutions upon non-Western cultures and environments[3]. It is only when we recognize that development is neither a neutral nor a universal process that we can begin to truly assess the ecological costs involved and begin to suggest some alternative approaches. If, as usually seems to be the case, Western values, concepts, and technologies become even more anti-ecological when they are transferred to or imposed upon cultures and environments to which they are not adapted, then any appropriate action by Western countries towards the poor countries would seem to involve either major changes in approach and/or high degrees of self-restraint in limiting the further spread of anti-ecological processes. However, since the Western countries—through colonialism—are responsible for many of the ecological maladies afflicting poor countries, such self-restraint cannot be equated with non-action; rather, it suggests quite different aid, development, and regulatory policies. Before these can be outlined, we must first elaborate upon some of the above points and try to assess the real ecological costs of current development approaches.

Ecological Costs of Current Development Approaches

Any attempt to estimate ecological costs—in whatever field— suffers from several fundamental difficulties. There is the inherent complexity of physical and biological processes, the understanding of which is made even more difficult by the fragmented and specialized nature of the academic disciplines. In addition, the conceptions and methodologies of those who specialize in costs—the economists—are in many ways

---

[3] In these terms, the USSR and most of Eastern Europe are included as Western countries. For a full discussion of the ways in which Western values are intermixed with modern science and technology, see Kenneth A Dahlberg, 'The Technological Ethic and the Spirit of International Relations', *International Studies Quarterly*, Vol. 17, No. 1 (March 1973), pp 55-88.

fundamentally anti-ecological.[4] Given these difficulties, the best that can be attempted here is a descriptive survey. Let us first look at some specific projects and their costs.

Specific Projects. It is important to include at least one historical project because many of the difficulties that flow from current development approaches relate to the indifference of most development economists to historical, much less ecological, factors. A recent careful study of public works project in 19th century India,[5] reads very much like many contemporary environmental 'horror' stories,[6] except that the historical remove delineates much more sharply the contrasting cultural conceptions and the very clear Western biases in what were then thought to be universal scientific principles.

The introduction of major public works programmes—in the form of canals, railroads, and roads—was expected to benefit both the peasants of India and the investors in England. Equally, the financial, administrative, and judicial reforms introduced after the demise of the East India Company were expected to remove the barriers to the natural workings of *laissez-faire* economics. The impact of these imported technologies, concepts, and institutions was, of course, quite different than expected. There were two serious effects. First, because the canals and railroads were designed with little concern for proper drainage, some 4,000 to 5,000 square miles of agricultural land was lost to salinity by 1891.[7] Second, while irrigation did increase production, it did so primarily for the spring export crops and at the expense of the autumn crops of millets and pulses upon which most of the population depended for their food and fodder. In addition to these 'natural' costs,

---

[4] They are fundamentally anti-ecological in the following ways: first, there is no attempt to conceptually link the economic system with real natural processes, say for example, energy transfers; next, economic theorizing is usually a-historical, attempting to ignore the evolution and real space-time context of both natural and social systems; finally, current economic thought contains a large number of value assumptions—especially those relating to growth—that are anti-ecological. For a detailed discussion and critique see Nicholas Georgescu-Roegen, *The Entropy Law and the Economic Process*, Cambridge: Harvard University Press, 1971.

[5] Elizabeth Whitcombe, *Agrarian Conditions in Northern India: The United Provinces under British Rule 1860-1900*, London: University of California Press, 1972.

[6] For perhaps the most complete catalogue to date, see J P Milton and F T Farvar (eds.), *The Careless Technology*: Natural History Press, 1972.

[7] Whitcombe, *op. cit.*, p. 11.

the canals also had the effect of increasing the wealth and power of those already well off. This was because the British left the construction and control of the minor channels and distributaries to the local rulers and landlords—who naturally used this power to increase their own wealth.

Large scale railroad and road construction compounded many of these difficulties. Drainage was reduced even more through the building up of these new embankments and barriers. Soil erosion and flooding was speeded up by the heavy demand for timber both for railway sleepers and fuel. Also, the deforestation of many areas meant that poor families now had to use cow dung for cooking fuel rather than as a fertilizer for their crops.[8] The introduction of British legal concepts of 'private property' and a series of courts to enforce its attendant rights, while not having an immediate impact upon the environment, did weaken traditional land tenure patterns by encouraging a spate of litigation over titles and debts. Thus, this 19th century attempt at development—albeit phrased in a different rhetoric, but with much the same mixture of idealism and interest as today—led to a serious deterioration of the natural environment, to severe social and economic distortions, and to an impoverishment of a large part of the peasant population.

Contemporary development projects are often planned by international civil servants rather than the colonial adminis- trators of old. However, the Western conceptions and approaches of the latter have been carried over virtually unchanged. This is most apparent in health, irrigation, and agricultural projects, where the approaches of the WHO and FAO—as well as those of various aid donating countries—have been basically technological. That is to say that various 'universal' technologies were promoted in an attempt to avoid the many specific cultural and political problems found in each region. For example, the use of DDT to attack malaria typifies Western ideas regarding efficient, inexpensive technological measures to solve a problem. The results show all of the dangers of specialized thinking, dependence on technological measures, and an attempt to avoid dealing with specific peoples in specific environments. One study shows that the levels of malarial

[8] *Ibid.*, pp. 93-94.

infection in Guatamala are now approaching pre-DDT levels while the mosquitos in the area are becoming more and more resistant to hard pesticides.[9] While the technocrat might argue that we have gained 20 years of protection, this is not the case. What has been lost is 20 years in trying to learn how to deal with malaria through biological/cultural means, while at the same time the tremendous long-term ecological costs of persistent pesticides have been imposed upon many tropical regions. What is now needed (and was equally feasible 20 years ago) are programmes that approach such matters from the village level. Various forms of vegetation management, the introduction of larvae-eating fish, the draining of shallow areas, the screening of huts, plus the use of selective biological measures such as the sterile male technique would go a long way towards reducing the dangers of malaria with little attendant environmental cost, while at the same time encouraging the sorts of village activities that most would include as part of any meaningful development programme. The barriers to such an approach lie not only in the vested interests of the urban elites in the developing countries, but in the lack of imagination on the part of national and international aid personnel. Also, such an approach requires not only imagination, but a great deal of *specific* information regarding the local environment, local customs, local patterns of political influence etc., so that the various possible programmes and technologies can be adapted to the requirements of the local situation.

Large scale dams, such as the Aswan Dam in Egypt, the Akosombo Dam in Ghana, and the various dams suggested in the Colombo Plan for the Mekong Basin, offer another example of large-scale projects, justified in the name of development, which have been conceived with very little concern for the larger environmental and ecological effects that they produce. The Aswan Dam, for example, while more than doubling Egypt's irrigated farm land, also greatly reduce the fisheries in the eastern end of the Mediterranean (the rich nutrients which formerly fed the fish now producing a rapid silting behind the dam and a profusion of plant growth in Lake Nasser). Ironically, the increased food production has been barely able to keep pace with the growing population—which itself is in

---

[9] F T Farvar, mimeo. paper presented to the AAAS Meeting, December 1971.

large part a result of cheap public health technologies—such as
water and sewage treatment, DDT spraying, and immuniza-
tions—which are much easier to introduce than birth control
techniques, since they encounter fewer cultural barriers. In
addition, the expansion of irrigation through canals had led to
great increases in the debilitating disease of bilharzia
(schistosomiasis)—a parasitic flat-worm that is carried by
microscopic snails living in the canals which enter the bare feet
of those working in the canals.[10]

The Akosombo Dam has created many similar environmental
problems. In addition, there were disruptions caused by the
dam flooding a number of tribal lands. While the government
made various studies in an attempt to minimize the difficulties,
and even offered to build new villages for those displaced, this
meant little to those tribes concerned because their cosmologies
were based on their living and dying, like their ancestors, on
that specific piece of earth. Not having our cultures so deeply
rooted in specific environments, it is easy for Westerns to
suggest that such displacement is a small price to pay for
progress—although the resistance of London homeowners to the
building of new motorways suggests that it is really the planners
and engineers who are insensitive to the real impacts of projects,
rather than those directly affected. It should also be pointed
out that while the Akosombo Dam does offer relatively cheap
electricity for urban elites and manufacturers, the major
beneficiary of the Dam is clearly the consortium of British and
American aluminium companies which obtained a guaranteed
long-term quota of electricity at fixed cheap rates with which to
process bauxite. While they did provide some of the financing
of the dam, they were also able to obtain significant financial
support from international agencies. It would appear that the
multi-national corporations are becoming quite adept at
obtaining international as well as national subsidies.

General Trends and Interactions

The above discussion of several specific projects suggests that if
environmental dimensions are disregarded there will be not only

---

[10] For a discussion of these and other health problems, both in Egypt and
especially in the Mekong basin, see J P Milton, 'Pollution, Public Health and
Nutritional Effects of Mekong Basin Hydro-Development', US AID mimeo. report
(n.d.).

unexpected ecological costs, but additional social and economic costs—which generally tend to fall most heavily upon the poor. A number of systematic dimensions are also hinted at and should now be made more explicit. The theoretical basis for this is found in the analysis of linkages between more and less organized subsystems made by the ecologist Ramon Margalef.[11] In discussing the links between a number of such subsystems, specifically including those between agrarian communities and industrial societies, he says that the latter subsystem . . .

. . . experiences more predictable changes through time. In so doing it stores information better and is a more efficient information channel. The first subsystem is subject to a stronger energy flow and, in fact, the second system feeds on the surplus of such energy. It is a basic property of nature, from the point of view of cybernetics, that any exchange between two systems of different information content does not result in a partition or equalizing of the information, but increases the difference. The system with more accumulated information becomes still richer from the exchange. Broadly speaking, the same principle is valid for persons and human organizations. . . . Such relations are compounded in an hierarchical organization and are reflected at every level.[12]

These relationships appear to be valid both at the national and international levels. Nationally, as industrialization has progressed, the farmers have generally become relatively poorer (on average—though there are similarly increasing gaps between rich and poor farmers). Internationally, the gap between the rich industrial countries and the poor agrarian countries continues to increase. While it is important to recognize such natural tendencies, in human affairs they are not to be accepted fatalistically. The basic purpose of welfare and development policies is presumably to place a limit on the size of the gap and to establish the levels below which society will not permit the poor to slide. Unfortunately, an unawareness of class biases at the national level and of general Western biases at the international level make many well-meaning policies counterproductive. In addition, many national and international programmes that are basically exploitative in nature are cloaked in the morally respectable rhetoric of welfare or development.

In terms of post World War II development approaches, two broad trends have been visible. First, there was the expectation

[11] Ramon Margalef, *Perspectives in Ecological Theory*, Chicago: University of Chicago Press, 1968.
[12] *Ibid.*, pp. 16-17.

that technical assistance and capital financing for industrial projects would be sufficient to launch the poor countries on the paths of modernization. This approach soon ran into serious cultural barriers—though it took development economists many years to recognize this. While the goals for the Second UN Development Decade include references to the need to deal with population pressures, to consider the employment consequences of new technologies, and to pay attention to pollution, the major thrust is still towards industrialization as the best approach to development. The development decades would appear to be a clear embodiment of the attempt to project Western values and technologies upon the rest of the world.[13] And if the ecologists are correct regarding the degree of change that will be required within the Western world, the development decades will tend to result in the promotion of obsolete approaches and technologies—that is to say, the West will be exporting its most anti-ecological products to the rest of the world at a time when it is changing its own approaches.

The second major trend, which is now gaining momentum, is the shift in thinking regarding agriculture that has occurred with the so-called green revolution. The biological idea behind the development of the new varieties of wheat and rice was that if tropical agriculture was to be improved, seeds specifically adapted to tropical conditions would have to be developed. Unfortunately, this partial ecological insight was vitiated by specialized and technological thinking.

Specialization led to a neglect of any consideration of the social and economic dimensions of peasant farming. Technological thinking led to the typical reaction once better seeds were produced: that they should be promoted universally, and more importantly, that peasant cultures should change—or be changed—to meet the requirements of this new and 'superior' technology. Peasants are thus expected to change cropping patterns, invest any capital they have in tube wells, fertilizers, and pesticides, and generally shift from local, barter-oriented

---

[13] 'The assumption that development is a *generalizable* concept must be seen in this context. It is far more potent than the crude instruments of "neo-colonialism". It is the last and brilliant effort of the white northern world to maintain its cultural dominance in perpetuity, against history, by the pretence that there is no alternative.' Address by John White, 'What is Development? and for Whom?', given at the Quaker Conference on 'Motive Force in Development', April 1972, p. 4. (My italics.)

cultures to larger, market-based systems. In short, they are expected to shift from what in most cases is an ecologically sound traditional agriculture to a form of modern industrial agriculture—with all its attendant ecological costs.

There are several ways of trying to estimate the ecological costs of such a shift. One would be to estimate the systematic impact of introducing high levels of both artificial fertilizers and pesticides. This has been attempted in at least three major—and quite different—surveys.[14] Each of these makes clear in its own way that industrial agriculture in the West is overdue for a thorough re-thinking and re-structuring. Another approach is to analyse what such a shift means in terms of global energy reserves (recognizing that our terrestrial stock of low entropy resources is limited):

... the ultimate and the most important result is a shift of the low entropy input from the solar to the terrestrial source. The ox or the water buffalo—which derive their mechanical power from the solar radiation caught by chlorophyll in photosynthesis—is replaced by the tractor—which is produced and operated with the aid of terrestrial low entropy. And the same goes for the shift from manure to artificial fertilizers. The upshot is that the mechanization of agriculture is a solution which, though inevitable in the present impasse, is anti-economical in the long run.[15]

A third approach is to consider the risks involved in planting huge acreages of crops of the same genetic stock. The spatial extension of single crops over large areas reduces natural variety and protection mechanisms—thus greatly increasing the risk from pests and plant diseases. Using crops with the same genetic base (hybrids) risks losing a large part of the crop if some new crop disease or mutation appears to which the crop is susceptible. This occurred recently in the United States when one-fifth of the maize crop was lost to a new variety of maize

[14] W H Mathews, F E Smith and E D Goldberg (eds.), *Man's Impact on Terrestrial and Oceanic Ecosystems*, London: MIT Press, 1971; The Institute of Ecology, *Man in the Living Environment*, Madison: University of Wisconsin Press, 1971: and D H Meadows, *et al., The Limits to Growth*, New York: Universe Books, 1972.

[15] Nicholas Georgescu-Roegen, 'Economics and Entropy', *The Ecologist*, Vol. 2, No. 7 (July, 1972), p. 17. Another dimension of this shift is that it involves moving from a highly localized form of production where there are high levels of re-cycling, to a process where many elements—particularly the fertilizers—are imported from outside, put through the crop, and 'exported' to rivers or ground waters, thus increasing the risks of eutrophication or contamination.

blight. Historically, the most disastrous example of dependence upon a single crop was the Irish potato famine.[16]

A fundamental misconception regarding the green revolution that clouds most discussions relating it to development is that it is seen as the only approach which offers sufficient increased production to offer a 'breathing space' in which to get population under control: an approach which—so the reasoning goes—must be adopted even with its great ecological risks. In fact, there are a number of alternative strategies, many of them much more ecologically sound, which offer the prospect of even *greater net production* than the green revolution. Recent research has suggested two important points: (a) that, for example in Iran, local improved varieties give better results than the 'miracle seeds' if they are given the same inputs;[17] (b) net production per unit of land on small farms—whether using traditional or new varieties—is greater than the equivalent crop (on similar soil) on large farms.[18] If this is so, one might then ask why the green revolution has been adopted rather than other approaches. One main reason is that in the real world of peasant farming the 'miracle seeds' are clearly 'landlord biased'. That is to say, that only the large landlords have the various resources—information, capital and/or borrowing power, and political power—to successfully adopt the new seeds and all of their requirements.[19]

Another main reason, which combines with the first, is that given the fact that in most developing countries a small number of people own most of the agricultural land, large increases in production result when these few landowners adopt a new technique that is more productive. What is argued here—and can

[16] Between 1700 and 1846, Ireland shifted from a grain-based agriculture to a potato-based one, with the population rising from 2 million to 8 million. During the years of famine, some 2 million starved to death and another 2 million emigrated. The inappropriateness of conventional scientific thinking at the time, the insensitivity of the British Government to the realities of the Irish situation and the actual aggravation of the crisis caused by bureaucratic centralization of relief measures and the dominance of *laissez-faire* economic thinking are all portrayed in detail in Cecil Woodham-Smith's, *The Great Hunger*, London: Hamish Hamilton, 1962.

[17] Ingrid Palmer, *Science and Agricultural Production*, UNRISD Report No. 72.8, Geneva, 1972, pp. 6-7. This would reduce genetic risks, if not those related to the use of irrigation and artificial fertilizers and pesticides.

[18] Keith Griffin, *The Green Revolution: An Economic Analysis*, UNRISD Report No. 72.6, Geneva, 1972, pp. 31-38.

[19] See *Ibid.*, pp. 46-48, for full discussion.

be seen in practice to a large extent in Taiwan—is that the same amount of land redistributed among a large number of small farmers would be much more productive.[20] Equally, such an approach would be labour intensive—drawing upon one easily available resource of the developing countries—and would not involve the degree of social and economic disruption that has occurred as the green revolution has increased gaps in wealth, led to greater unemployment, and often reduced the local high protein crops the poor depend upon.

It is instructive to consider the sort of agricultural research that would be required—in addition to land redistribution—to make peasant agriculture more productive while maintaining its ecological soundness. The research would have to be self-consciously designed to be 'peasant biased', that is to say, it would have to consider how to improve local food crops, how to improve varieties that are grown on rain fed, rather than irrigated land, and how to reduce the losses to local pests, etc.[21] Such research would involve a massive research investment in each country—because to be ecologically sound, new varieties, husbandry practices, protective measures, etc, would all have to be adapted to the local ecological and cultural conditions. And, of course, the research would have to be interdisciplinary to include the social as well as the physical dimensions of agriculture.

Such a research and aid programme—which would require a major shift in attitudes on the part of both government and international agencies—would tend to reduce another major risk in the green revolution. That is, that as the multinational corporations become more involved in the commercial aspects of the green revolution—the selling of seeds, fertilizers, pesticides, tractors, irrigation equipment, etc.—they will tend to want to standardize (and limit) the kinds of seed available, the number of different mixtures of fertilizers, the size of tractors, and so on. What this really means is that the peasant becomes much more dependent upon outsiders. Decisions as to what the

---

[20] What often makes the large farms appear more productive than they are is the tendency to measure productivity in terms of production per man hour or in capital terms rather than in terms of crop productivity per land unit.

[21] Given the fact that crop losses in the tropics average 25-45% of the crop (from the field through milling), Ingrid Palmer, *op. cit.,* p. 72, asks what might have been the result had research since World War II been directed at reducing these losses, rather than concentrating on increased production through hybrid seeds.

appropriate seeds and fertilizers for him will be are more likely
to be made on the basis of ease of production and commercial
profit than upon ecological grounds that relate to his region and
his farm. At the same time the new cultivation practices force
him to abandon many traditional ways of dealing with pests,
etc.—with the clear danger that his localized knowledge will
soon be lost. These points show the great importance of
including the actions of private corporations and investors when
considering the 'developmental' impact of Western agencies
upon non-Western peoples.

Conclusions

The above discussion hopefully makes clear the need for new
conceptual and organizational approaches to understanding and
dealing with 'developmental' and environmental dilemmas.
Conceptually, three easily stated, but most difficult to realize
changes are needed. First, there is a need for much less
'universal' thinking and categorizing (which as mentioned often
contains large admixtures of Western values) and much more
'contextual' thinking—thinking that relates to specific groups in
their real natural and historical environments. Next, and related
to the first point, is the need to recognize that technologies are
not neutral, but reflect (as well as shape) the values and
environments in which they were developed. Once this is
recognized, then perhaps we can go about the more important
task of learning how one *adapts* technologies to fit different
cultures and environments (this would suggest that we should
ban the phrase 'transfer of technology'—since such 'transfer' can
only occur through imposition). Finally, it is clear that
conventional economic theories and measures are inadequate, if
not positively misleading, in developing ecologically sound
policies—which as has been stressed, often are more productive
in real economic and social terms. When all the above points are
taken into account, much of the supposed 'conflict' between
environmental and developmental goals disappears.

The sorts of organizational changes which are suggested
include the following. To the degree that international
development programmes are encouraged, they should be
conceived in terms of counteracting and placing limits on the

natural tendency for resources and information to be extracted by the richer, more organized subsystems from the poorer, less organized ones. This, of course, would require a thorough re-thinking of current approaches to aid, trade, and the activities of multi-national corporations. In terms of specific projects there would appear to be a general need to research and develop low-level, peasant-biased technologies that are adapted to the particular culture and environment concerned. While some of this might be done on a fairly broad basis, there is also a great need to have specific projects—such as dams, canal systems, etc.—planned from the beginning by interdisciplinary teams that include some form of real participation by the various groups affected by the project. The fact that such planning and participation is seldom realized (and usually feared) even in most Western countries suggests both the arrogance of most 'experts' and the great distance we have to go to learn how to adapt and humanize our own technologies, much less reduce their harmful impacts on other cultures.

Addendum

A conference like this should stimulate us all to re-examine our basic assumptions and to question the various optimisms and pessimisms that underlie our discussions. Let me add some sharpness to some of the general points that have been made so far. A great deal has been said regarding the need to think of the qualitative aspects of life as well as of the quantitative ones. One way of starting this is to re-work various quantitative measures to include various ecological dimensions that have been left out. For example, with one exception that I can think of (Georg Borgstrom), none of the population statistics used to estimate global food needs include the food requirements of livestock and wildlife! And though people may talk a great deal about the weather, both historians and futurologists tend to project current climate patterns back into the past and into the future. They do this in spite of the fact that there is clear evidence that there have been major changes in climate over the past centuries and that we may be undergoing a significant secular change at present. What happens to global projections should the Soviet winter wheat crop fail for lack of adequate snow cover the next 10 out of 20 years? Shouldn't there be much more research on these sorts of long term climatic changes?

At the other end of the time scale, there is now considerable research on circadian rhythms in plants, insects, and animals (daily cycles or changes in temperature, hormone levels, etc.). Among a host of interesting implications, let me mention only one: that it has been found that a particular type of flour beetle is twice as susceptible to pesticides at 2 am in the morning than 2 pm. The prospect of being able to reduce greatly the amount of pesticides required for a particular effect simply by proper timing would seem to be worthy of a great deal of research; however, my impression is that most of us continue to assume a similar functioning of plants, insects, and animals at all times of the day. And, of course, the thought that production lines should be keyed to bodily rhythms or that the same sample survey run at different times of the same day might give different results would send shudders down the backs of both industrialists and sociologists.

While a great deal can be done to make quantitative measures more sensitive to environmental dimensions, it must be recognized that the 'ecological debate' ultimately revolves around the problem of how society can gradually place a run-away technological-economic complex back under the control of a larger value framework. The debates about 'progress', 'growth', and 'quality of life' all involve an attempt to suggest that human values and needs must have precedence over the requirements of technology. It is here that underlying moods of optimism or pessimism take on great importance. I find it most interesting that our responses to threats of disaster vary so greatly. On the one hand, our response to the threat of nuclear war—a real, but presumably limited and controllable threat—is one that involves extreme examples of what is called 'possibilistic thinking', i.e. building defences and deterrents against the worst possible weapons that you can imagine your opponent possessing. On the other hand, when responding to the threat of environmental catastrophe—which is both less obvious and less controllable—most people firmly reject any sort of possibilistic thinking. In fact, some may even reject the threat itself, a demonstration of some sort of backlash to the environmental movement. Even if a threat is recognized, most economists and industrialists engage in what I would call technological optimism—in effect saying that because we have muddled through in the past, we can count on technology and good sense to carry us through any future difficulties. Why is the advice of the military strategist so easily accepted—even when his 'defence systems' may cost billions of dollars—when the advice of the ecologist, who is also trying to defend and conserve, is received so coldly? Certainly, there ought to be major re-allocations between the defence budget (military) and the environmental defence budget, but this would not necessarily influence trade and aid in favour of the developing countries. Which leads to a final point.

In talking about aid and development, we have talked mostly about administrative dimensions—the procedures for planning projects, better ways of utilizing existing resources and personnel, the pros and cons of national versus multinational approaches, and so on. There has been some discussion of what might be called the legislative dimensions of development, such as the United Nations Development Decades, the various attempts to persuade the rich countries to increase the percentage of their GDP that is allocated to aid, etc. What has only been touched upon is what I would call the constitutional level of aid—the basic framework within which aid and trade is carried on—and which will have to be changed if there is to be any real shift in resources from rich to poor. It is important to recognize that changing the terms of trade between rich and poor—as has been mentioned, the UNCTAD conferences tried unsuccessfully to do—would probably be of much greater benefit to the developing countries than an increase in aid. In this regard, I am surprised that nobody has mentioned the Law of the Sea Conference scheduled for 1974. This will be a major 'constitutional' conference, establishing the basic frame-work for years to come of who will be able to exploit which ocean resources for whose benefit and under what sort of regulation (if any). We must remember that we are talking about some two-thirds of the world's surface. So far, it seems that the Law of the Sea Conference is going to be dominated by the views of international lawyers on the one hand and those of the various high technology interests on the other—including both the military establishments and the oil drilling and manganese nodule mining groups. While there will be competition between these groups, it will only be in terms of how the oceans will be carved up for exploitation. Few people have discussed the implications of this conference for environmental protection. And while the developing countries sense some of the importance of this conference for their development prospects, they know that as with the UNCTAD conference, it will be very difficult for them to counter the power and position of the industrialized countries. By introducing the environmental element, you both strengthen the arguing position of the developing countries and you question some of the conventional standards for measuring utility, as they relate to both 'development' and to the value and use of the oceans. The question which then follows is which groups, if any, are going to make a concerted effort to introduce such dimensions into this more important conference?

Discussion

OWEN JONES

I am greatly in sympathy with the underlying theme of Professor Dahlberg's stimulating paper, but I would like to raise one or two points. As an economist much concerned with environmental matters, I do not find the sort of conflict which Professor Dahlberg suggests. The economist is concerned with maximizing human benefit from scarce resources which can be used in alternative ways. This should not clash with Human Ecology and many of the analytical tools developed and used by the economist are of essential value to the ecologist as well; so these two activities may not be at variance.

Secondly, a more specific point: the 'demon DDT'. It is natural to have some disquiet about the persistence of this substance—I gather that it has been found in the fat of penguins in the Antarctic, so we all tend to say that this is absolutely terrible—but in fact the penguins don't mind! I have searched hard and have failed to find any convincing evidence of the harmfulness of DDT to man and have found very little on the harmfulness to other species apart from intended victims. Of course I accept Professor Newbould's paper this afternoon with its reference to certain birds. I also accept his figures for the algae off the coast of Monterey, which I interpret as a 'long distance warning'—to be taken seriously but hardly at this stage a danger signal.

On the other side of the equation, apart from the enormous increase in food and fibre production which DDT has allowed, more than 1000 m people in the past 25 years have been freed from the risk of malaria (World Health Organization figures) almost entirely through the use of DDT. Even at the present day about 300 m people rely on DDT for their protection against this really dreadful human scourge.

DDT is being phased out, and probably rightly, but we have

to weigh these two things in the balance—the long-term possible ill-effects of DDT, and the enormous advantages which have accrued and are still accruing from its use. Personally I am prepared to accept this particular risk. Biological control has been suggested as an alternative, but in the situations in which DDT has been used, this is not a practical proposition. I doubt whether the necessary resources could have been mobilized, even if the techniques were available (and I doubt if they were) to secure the effects that DDT had at this stage. Also we must beware of looking at biological control as something pure and lily-white as opposed to the filthy chemicals. In fact anyone who has watched rabbits dying of myxamatosis will plump for cyanide any time on humanitarian grounds, and this is quoted as one of the most unambiguous and effective example of biological control. Since biological agents are subject to mutation they may one day work in a completely different manner from that intended. Some of the 'controlling agents' introduced to control pests can in due course become worse pests themselves. Biological control does have its place, but is is by no means the complete answer, and there are occasions when chemical control must be considered.

Thirdly, the Green Revolution. Of course there will be the kind of problems which Professor Dahlberg mentioned. But I was in India in 1958 and saw the depressing situation existing at that time with the country dependent for its day to day survival on American food aid, and no foreseeable improvement. In 1965 I revisited India and the whole climate of opinion in the country had changed. Where before there was despair, now there was hope. I saw, at first hand, the early effects of the new cereal varieties, the start of the Green Revolution and the transformation of attitude. In physical terms, it meant that the previous maximum wheat production of about 12 million tons around 1964 was eventually increased to double that level by 1971. In political terms, it meant much greater self-reliance for India at least in terms of essential foodstuffs, and the restoration of self-esteem. These are the factors which we have to weigh against the unfortunate side effects of the green revolution and personally I have no doubt where the balance of advantage lies.

We have reached the point where the world is aware of the ecological problems, and we owe a debt to those who, by their enthusiasm and in spite, sometimes, of exaggerations and half-truths, have made this so. Now we have to devote our efforts to reaching a fuller understanding as the basis of more effective management of our environment—efforts which will require the co-operation of all the disciplines that make up Human Ecology.

KENNETH DAHLBERG

I don't think we disagree as much as may appear. But there are several points I would like to take up. First, the role of economists. They would be of much greater help if they could concentrate more on the economic costs of environmental impacts, and devote attention to trying to develop ways of measuring environmental impacts in economic terms.

On the question of DDT, I didn't want to suggest that it should be abandoned entirely, but to remind people of the tremendous problems which have arisen in countries such as Guatemala as a result of the development of resistance to DDT in insects including the malaria mosquito. I agree it is very hard to assess damage which DDT does to other living organisms including man, but what does worry me are the possible long term effects. For both these reasons, I see a much stronger case for integrated control measures, using a variety of approaches, including, possibly, selective use of chemicals. When DDT first came in, we thought it would be of permanent value. Now we are finding many different problems which we were totally unable to forsee at first.

Concerning the Green Revolution, we seem to be moving in a pattern first of despair followed by hope, and I think we may be now moving back towards an outlook of despair, partly at least because we paid insufficient attention to the social and other effects produced by the use of the new crop varieties, some of which I have dealt with in my paper. I am particularly concerned with how one gets from statistical averages, improvements in yield in field trials, to improvements in the quality of life of village groups starting to use the new varieties. Because of their agronomic demands, the varieties can often accentuate differences between regions and also between rich and poor farmers. I am particularly sceptical of claims of development based on national averages. All too often, these cover gross inequalities within the country concerned.

One further criticism which I have of many development approaches is their lack of concern with real processes of education. All too often, this means training for one particular and restricted function, not an attempt to show a person how to realize his total capabilities. On a national scale this also involves an increase in research and development undertaken in and by developing countries. At present less than 1% of world research and development takes place in developing countries and until they acquire independent capacities for research and development they are going to be dependent on developed countries.

MILES DANBY
Concerning your comments on public works programmes, I would not disagree with them but would like to add some of the things you did not say, as I think there is an implication that major public works schemes are 'out', whereas 10 years ago they were 'in', rather like national airlines.

This would be wrong as there are many large public works schemes which have been in operation for many years which have worked very well. One of the most obvious examples is the Gezira scheme in the Sudan, under development since 1920, with very little in the way of ecological side-effects and a scheme of great importance to the Sudan.

Concerning your remarks about the Akosombo Dam, I was under the impression that the resettlement involved had been achieved relatively satisfactorily, and that it is certainly too early to consider this a failure. Also, the idea of the Volta Dam dates right back to about 1915 and several schemes were developed over the years before the adoption of the final one, produced by Kaiser Aluminium and supported by President Nkrumah. The Ghana government contributed a major part of the cost, so I do not think this was a case of outside interests imposing this on the country concerned.

KENNETH DAHLBERG
I didn't mean to imply that there is no requirement for large scale public works, but more that they must be planned with careful attention to environmental impact. In addition, more attention should be paid to having a number of small projects in place of one large project. What I would challenge is the

frequent assumption that the big project is necessarily the best.

Although accepting your point about Akosombo, there have, without doubt, been many examples of disastrous environmental and social impacts of major programmes, too often resulting from the utilization of methods alien to the country concerned.

On a general point, my intention is to point out some of the disadvantages of major development projects. I fully appreciate the benefits which accrue, and have assumed that we are all aware of these but not quite so aware of the drawbacks.

Ecological Approaches to Agricultural Development

Jonathan Holliman

Introduction

The purpose of agriculture is to convert solar energy, various compounds, gases and water into a product that is useful to man and which can also be made available later than the time of production and in another place. Its most important product is food although plant and animal materials serve many other purposes. The production of food for the world's human population by agriculture, hunting and gathering or by industrial processing is the main concern of this paper.

The development of agriculture has involved a series of innovations nearly all of which have been concerned with manipulating ecological processes so as to increase the proportion of total biomass that is useful to man as well as the rate at which it is produced.

There is general agreement that the beginnings of agriculture occurred in the Middle East about 7500 BC and were based on the cultivation of wild species of wheat and barley and later the herding and domestication of cattle, sheep, goats, pigs and other animals. This process of domesticating wild plants and animals and the dispersal of seeds and stock from one continent to another is one which has become more and more sophisticated as techniques of genetic engineering have improved.

Other agricultural innovations were concerned with clearing land for cultivation by fire or mechanical means. The harnessing of animals for work extended the energies of man yet did not initially compete with him for food. The introduction of irrigation in the Nile and Tigris and Euphrates valleys about 6,000 years ago created the first large scale food surpluses allowing the major expansion of urbanisation, division of labour and the pursuit of non-food producing activities. In the past two centuries, scientific and technological advances have produced the most radical changes in agriculture. Mechanical power driven by mineral fossil fuels, the mining and production of mineral fertilizers and the use of chemicals to control pests, weeds and diseases have permitted man to manipulate the ecosystem more extensively than ever before.

Population/Food Equation

All these agricultural innovations have increased the amount of food available to man's expanding population. More people could produce more food to feed yet more people. This mutually interdependent increase in the crop of both food and humans has recently become out of balance on a major scale. This has been largely due to the abandonment of social controls on population and the rapid decrease in infant mortality made possible through the widespread adoption of pharmaceutical and sanitary techniques. In many areas of the world, population is expanding at a faster rate than food production. Of the present world population of 3.7 billion it is estimated that approximately one third are continually hungry and receive less than the average daily required minimum of 2,700-3,500 calories. A further third may not be receiving sufficient proteins or vitamins and are therefore malnourished.

The individual nutritional requirements of humans and the size of the gap between what is available and what is the basic need must therefore be the starting point for any consideration of the prospects for alleviating or solving a food shortage which at this moment seriously affects over 2,000 million people. Unfortunately these nutritional requirements are difficult to determine precisely. Is it enough to provide a diet which just prevents disease or must we supply sufficient for optimum health? The minimum amounts to prevent disease and the maximum which will continue to contribute to health and growth are not known for all nutrients. This is complicated by the interactions between nutrients. An increase in fatty acids intake will increase vitamin E requirements; protein may be sufficient but if calorie intake is too low then proteins will be metabolized by the body for their calorific content. There is also an obvious difference between individuals in size, sex and age as well as the amount and type of work they do and their susceptibility to disease. The timing of nutritional intake is important for children and pregnant mothers where deficiency can cause permanent physical disability and retardation of mental development.

Thus, in planning for the needs of whole populations, the varied individual requirements must be allowed for. For some nutrients it may be dangerous to assume that the average intake would be a sufficient guide. The consequences of some people getting less vitamin or protein than they need, even though this may be the average requirement, may be much more disastrous than if they received the average calorie requirement since activity can be more easily adjusted to calorie intake whereas physiological processes cannot be adjusted in the same way. Clearly the problem of distributing the right food to the right people at the right time and allowing for margins of safety is just as important as being able to meet the total average food requirements. The cultural acceptance of particular foods must be another consideration. Bringing some variety into the incredible monotony of most diets will put further limitations on food supply. Minimal requirements are not enough when people quite naturally aspire to a varied and tasty diet.

Despite these problems of assessing individual food requirements we know that, according to the FAO, a population growth of 2.6% a year will require a growth in food production of 3.9% per year just to provide a minimal diet. Since the actual growth rate of food production is 2.7% at the moment, the difference between the two represents the amount that needs to be made up. The various efforts to meet this shortfall constitute the main components of present agricultural development. Any ecological approach to this development process must consider the limitations on the general availability and rate of supply of materials and energy necessary for plant and animal growth, take into account the impact of agriculture on the non-food producing parts of the ecosystem, and assess the actual ecological efficiency of the production that is achieved.

Efficiency of Energy Conversion

The basic 'currency' of any ecosystem is the energy available to it. The efficiency of biomass production can be measured in terms of this energy currency and it also provides a universal standard by which the productivity and efficiency of different ecosystems and different agricultural systems can be compared.

Solar energy enters the atmosphere at a constant rate but due to the curvature of the earth's surface, its orbital path round the sun and climatic factors such as cloudiness, the energy reaching the surface varies from about 200 kilocalories per square centimetre per year in desert areas to 70 kcals/cm^2/year in the polar regions. Tropical areas receive from 120-160 kilocalories and temperate Europe 80 to 120. At the equator there is very little seasonal or diurnal variation in solar energy received over the year but due to increasing day length at higher latitudes, areas such as southern Scandinavia have the highest potential for photosynthesis during the summer growing season. Expressed in grams of net plant productivity per square metre per day, potential yields for these high latitude regions are up to 38 gms while for many tropical areas the maximum is only 27 grams. This means that nearly all so-called developing countries are limited climatically to much lower potential rates of photosynthesis than countries in temperate regions [1]. However hard they try, yields of rice with a four month growing season in tropical countries could never be expanded to reach those of Spain for instance. This can only be overcome by growing two crops in a year or by crops such as sugar cane which have a much longer growing season and can take advantage of the fairly constant daily incidence of solar radiation throughout the year. Obviously, natural ecosystems with complete plant cover throughout the year may achieve higher yields than agricultural crops spaced out and grown for only part of the year even though the crop plants themselves may be more efficient converters of solar energy.

In various parts of the world, potential photosynthesis is limited by cold, water supply and nutrients rather than incidence of solar energy. Where these other limitations are overcome by man, crop production will substantially exceed ecosystem productivity under natural conditions.

Aside from overcoming limitations on inputs, agricultural development has been a process of simplifying natural ecosystems so that net productivity can be channelled into crops for humans rather than sustain the cyclical processes and intricate relationships and control mechanisms of a climax community. Even though this process, by itself, may decrease total productivity of an area it still increases the proportion of the remaining yield that is of direct use to man. But when this is done the additional energy required to manage the agricultural system, for controlling weeds and pests, for sustaining material inputs, and for harvesting and transporting the crop, must come from human and animal effort or from energy stored in the form of fossil fuels.

In a pre-agricultural system or hunting and gathering society, manipulation of the natural ecosystem is minimal. Pygmies in tropical rainforests take a small part of the yield from each of many types of fruit and animals available. Population densities are low and productivity can be maintained due to the complexity and stability of the ecosystem. In ecosystems where annual productivity is lower or there is seasonal variability, man has made certain adaptations. Eskimos survive by cropping the productivity of extensive areas of aquatic ecosystem and by migration from the water's edge in winter to summer land areas when wild plant foods are also available. Where water is the main limiting factor desert bushmen may range over hundreds of miles to hunt for food. In temperate regions with seasonal fluctuations, hunters such as the American Indians of the prairies would follow the migratory routes of the Buffalo. In all these systems solar energy is supplemented by small amounts of energy from human work or from domesticated animals such as husky dogs and horses, which are used for transport when migration is necessary. An intimate knowledge of species and their habitats, seasonal cycles, etc., is essential in order to crop the natural ecosystem in the most efficient way.

With simple agricultural systems man has just gone one stage further by using certain domesticated animals and plants in place of wild species. They are still subject to seasonal changes and variation in inputs such as water but the plants and animals act as a controllable storage system making food available during unfavourable periods. Draft animals are used to do the planting, fertilizer spreading, weeding and water pumping while still acting as a walking larder and protein source. Nomadic systems based usually on domesticated cattle or sheep are closely adapted to seasonal productivity and year after year follow defined routes from pasture to pasture. So long as population densities remain stabilized below the carrying capacity, nomadism may be a very efficient form of agriculture on poorer rangelands.

Where grain crops replaced natural grasslands, agriculture followed the same seasonal pattern; but where complex, stable and perennial tropical rainforest is replaced by simple, annual cereals, agriculture developed differently. Once the forest is cleared, solar energy is then available for crop production. But this system has lost the features of the climax forest; biological diversity, control of epidemics of single species and main‑ tenance of soil fertility. Thus pests and weeds invade the crops, and nutrients built up by the forests are gradually dissipated. After one to three harvests the cultivated area must be abandoned. Thus without the additional energy required to artificially maintain nutrient supplies and control pests and weeds the farmer has no alternative but to move elsewhere and start again. This practice of shifting cultivation is widespread in all tropical forest areas. Although it may appear to be more productive than a relatively sedentary hunting and gathering system some people consider that crop yields from one or two years for every 20-50 years of fallow period may in fact have the same order of population carrying capacity.

The availability of concentrated inputs of power from fossil fuels from the beginning of the last century has led to the industrialization of food production. Crops are sown and harvested and the soil prepared using machinery run on fossil fuels, weeds and pests are controlled by chemicals derived partly from fossil fuels and processed with energy also derived from them and water and nutrients are also processed and transported with the aid of energy from these sources. Fossil fuels also support the research and managerial system required to maintain the flow of inputs and crop harvests. Yields rose as human and animal energies were replaced by mineral energy sources and the number employed in agriculture has declined proportionally and shifted from the land to the laboratories and fertilizer factories. A higher proportion of the calorific value of crop yields now comes from the additional mineral energy inputs rather than from solar energy. Food under industrialized agricultural systems is therefore partly made of fossil fuels.

By supplementing food production with fossil fuels, industrialized agriculture gives the impression of being an efficient converter of energy but although industrialized agricultural yields per unit area may be much higher than yields under shifting agriculture the return on the additional energy investment may be small (the conversion of solar energy remains more or less the same). For the Tsembaga of New Guinea their shifting agricultural system gives more than a 16 : 1 return on the additional human energy investment required for clearing, weeding etc.[2] For a plant crop under a modern agricultural system the ratio of additional fossil fuel input to calorie yield may be only 3 : 5 [3]; with nomadic cattle and intensive beef cattle production the comparison is even more unfavourable. For range cattle a solar energy derived input of 100 kcals/m^2/day yields 1 kcal of meat. For an intensive beef cattle system of pasture and fodder 100 kcals/m^2/day of solar energy derived input with an additional 50 kcals/m^2/day from fossil fuel sources yields 40 kcals of beef. The ratio of additional input to yield is 5 : 4. It is an almost direct conversion from fossil fuels to meat calories. An industrialized food production system based on fossil fuels will obviously only be viable for as long as the additional mineral fuels are available. When these become scarce we must return to a solar energy based economy.

Efficiency of Protein Conversion

Concentration of agriculture on a few crop species has meant that protein content has been reduced relative to that in mixed diets derived from hunting and gathering. The new tropical varieties of rice and wheat highly responsive to fertilizer and water inputs also yield 10-25% less protein. Man can thus increase the range of plant food crops to supply protein or eat meat and fish products where the necessary protein has already been synthesized by other animals. A further alternative is to produce protein direct from fossil fuels by chemical or microbial processes and so by-pass higher plants and animals. Although the techniques are developed, fossil fuels are again a possible limiting factor on maintaining production indefinitely.

It is a well known rule-of-thumb in ecological energetics that at each trophic level from primary producers to herbivores and then carnivores, the conversion ratio from the lower level to the next is 10 : 1. This means that 1,000 kcals of plant food will maintain 100 kcals of herbivore which can be converted into only 10 kcals of human biomass. Thus the energy base for meat protein supply is much larger than that required for plant protein production. The lower the trophic level at which protein is cropped then the greater the efficiency of solar energy use.

Animal species, although converting plant food at the ratio of 10%, have different rates at which the conversion can take place. This is due to differences in biomass, metabolic rates and breeding rates. Thus one cow may gain 240 lb from 1 ton of food in 120 days whilst 300 rabbits with a weight equal to that of the cow will convert 1 ton of food into 240 lb of meat in only 30 days [4]. In terms of meat production per unit of time rabbits are four times more efficient than beef. Other factors to take into account are the ratio of edible muscle to non-edible tissues and losses from infant mortality. Pigs also have an advantage in that their food is often the 'waste' from food processing and consumption. It is also known that wild range animals such as the antelope and zebra are much more efficient than domesticated cattle at producing meat from African savannahs. They can eat a wider range of plant food and are also more resistant to disease such as Tsetse fly. Because fish do not need to spend energy on keeping warm they are also considered to be very efficient converters of plants to meat protein. By the judicious use of selected species, efficiency in protein production can be improved but it will not compare favourably with the efficiencies possible from the direct consumption of plant protein.

Efficiency of Land Use

Food production systems based on solar energy alone require a wide range of species and a certain amount of land. Hunters and gatherers in tropical rainforest can each be supported by 1 square mile of forest. In a subsistence agricultural system with domesticated animals and plants such as that in India, densities of 640 per square mile are sustainable. In Britain it has been calculated that each person is using 1.4 acres to supply our varied diet, most of the land being outside its territory. It is well known that peasant agriculture is a more efficient user of land area than modern industrialized agricultural systems. A ranking of various countries for yield per hectare and per active male in agriculture shows Taiwan at the top of the list followed by the

United Arab Republic [5]. Smaller units can also be shown to be more efficient. In Cornwall the average output for holdings of less than 50 acres is twice that of the national average In fact, for all types of farming in Britain yields decline as acreage of farm units increase [6]. It seems that once agriculture becomes part of an industrial system, ecological and resource efficiency rules are exchanged for economic ones. The Rowatt Institute in Britain has analysed the relative efficiency of beef production from an intensive barley feed system and a grass system with some seasonal outdoor grazing [7]. With the intensive cereal feeding system where cattle are kept indoors, gains of 1.2 kg per day and feed conversions of 6 kg barley per 1 kg of live weight gained can be achieved. With high quality grass which is artificially dried and fed to cattle when they are not grazing, overall gains of 0.85 kg per day and a feed conversion ratio of 10 kg of dried grass per 1 kg of liveweight gained are achieved. Barley beef looks a better proposition on this basis but the per acre production of dried feed from grass is three times that from barley. Therefore there is nearly twice as much gain *per acre* from grass production as from barley grain production and the basic land input is used more efficiently. But dried grass production requires more capital investment, the beef gains are less rapid and cattle are slaughtered at heavier weights. This means that the rate of fattening is slower and more costly labour is required. The barley beef system is adopted not because it uses land more efficiently but because returns on capital investment are more favourable.

Since capitalist economics do not relate closely to efficiency in resource use and since future costs, depletion of resources and environmental side-effects are usually discounted, relatively short-term profitability can continue even though ecological efficiency and eventual diminishing returns clearly show that potential crop productivity is not being maximized. In general, profitability in modern industrialized agriculture is more important than sustainability or efficiency.

Limitations on Land and Soil

Land represents the surface area where solar energy is gathered and the point where the main interchange between organisms and their nutrient and water requirements takes place. Just below the surface is the soil, a complex ecosystem in itself which plays an essential role in recycling nutrients and regulating their supply. Although extra land surface cannot be created on any significant scale the variable rate of inputs to land surface can be adjusted to make the available land surface more suitable for agriculture yet, according to the FAO, the most easily cultivated land in the world is already under agriculture. It will also cost approximately $28 billion a year to develop enough new land to feed the annual growth in world population [8].

The potentially farmable area of land in the world has been estimated to be 7.86 billion acres, almost three times the area at present in use in any one year [8]. But this potential land must first be cleared and a large proportion of it will require additional inputs, especially of water. The apparent condition and fertility of the soil may also be mythical. As experience has constantly shown, clearance of vegetation in tropical areas can lead to laterization in a matter of 20-50 years. Tropical soils are not necessarily fertile. It is the high rate of humus accumulation and decay which gives the appearance of fertility.

The soil structure may also be affected by changes in land use. Complex relationships between micro-organisms which contribute to fertility may be broken. Physical impaction by cattle and heavy machinery can impede soil drainage and lead to waterlogging and disappearance of soil organisms. The destruction and erosion of the soil by over-grazing and other agricultural practices is estimated by Russian scientists to be destroying land at a faster rate than new land is being made available by irrigation. A French scientist, Alexis Guerin, also estimates that during the past century wind and water erosion has destroyed about 5,000 million acres, equivalent to 27% of the world's land in active agricultural use [9].

Limitations on Water Supply

Water is the essential input that will increase land availability to the greatest extent. Although precipitation is highly variable in its areal distribution the rate at which precipitation follows evaporation in the hydrological cycle still occurs at a constant average of 9-10 days. The variability in distribution means that precipitation must be stored and channelled to make it available at the right time and in the right place for optimum plant and animal productivity. The side effects of increased water supply such as waterlogging and salination, and spread of water borne pathogens such as bilharzia, must also be taken into account as part of the cost of doing this. Even then, overall freshwater supply is a serious limitation on future food production; George Borgstrom, Professor of Food Science at Michigan State University, considers that drastic world shortages of water will result in mass starvation within the next few years [10].

Salt water irrigation in desert areas is still in its early stages but may alleviate many of the supply and distribution problems and side effects associated with freshwater irrigation for the arid tropics. The cultivation of halophytic crops is also receiving urgent attention.

Limitations on Nutrient Supplies

After water, the next major input requirements for agriculture are mineral nutrients—mainly nitrogen, phosphorus and potassium. In natural ecosystems these and other minerals are involved in cyclical processes which constantly replenish the supply from the decomposition of dead organic matter as well as other sources. The rates at which these cyclical processes operate limit the ecosystem productivity. More intensive crop production therefore necessitates the artificial replenishment of the nutrient supply.

By far the largest part of the nitrogen available to crop plants comes from the fixation of atmospheric nitrogen by bacteria in the roots of leguminous plants and by some other soil micro-organisms and algae. By planting legumes subsequent to other crops and ploughing them into the soil, its nitrogen content can be increased. The other method is to produce ammonia from pressurized gaseous nitrogen and hydrogen by the Haber process. Artificially produced ammonium compounds can be spread on the soil and are then converted by soil organisms into nitrate, the form in which it is taken up by plants.

Phosphorus fertilizer is mainly mined as calcium phosphate, most of which was deposited on the sea floor about 400 million years ago. The calcium phosphate is converted by heating or by the addition of sulphuric acid to make superphosphate, containing a proportion of phosphorus pentoxide, which, being soluble, is more easily available to plants. There is no immediate shortage of phosphate rock and, given enough energy, the supply of nitrogen and hydrogen for ammonia products presents no problem (each ton of nitrogen fertilizer is said to require 5 tons of coal or its equivalent to produce). Potassium compounds are also mined and processed to make fertilizer and at the present rate of consumption known reserves are likely to last for at least a further 600 years.

Although chemical fertilizers have considerable advantages over organic sources and supplies are reasonably assured for the future, the increasing quantities applied to soils are not without repercussions on other parts of the biosphere. For example, the condition of metahaemoglobinaemia caused by the assimilation of nitrites which have entered water supplies has led to the death of cattle, sheep, humans and wildlife. The inadvertent fertilization of enclosed or slow-flowing aquatic ecosystems, especially by phosphates, ensures that the stage of eutrophication is reached very much sooner than it would under natural conditions. Supplies of fish and other useful aquatic organisms may be eliminated as a result. There are other dangers in using large concentrated doses of fertilizer which produce changes in soil chemistry and may alter adversely the nutrient balance in the plant crops themselves. As a result, diminishing returns on inputs will occur. An 11% increase in agricultural production in the US between 1949-68 was achieved with a 648% increase in nitrogen fertilizer while an 800% increase in nitrogen fertilizer was necessary for a 35% increase for Britain during the same period.

Limitations of Genetic Resources

Whatever means of manipulating the supply of water and nutrients is used, the chosen plant will only grow to a certain size and once the maximum is reached no amount of further inputs will result in higher productivity. The limits to individual plant growth when all other limiting factors are removed are determined by its genetic characteristics. To increase crop production a more efficient variety with a different genetic composition must be found. Genetic experimentation has developed thousands of varieties of the common cereal crops which can provide the maximum yield under a specific set of environmental conditions. The most important result of recent developments in plant breeding has been the introduction of tropical varieties of wheat, rice and maize, which are highly responsive to inputs of water and the application of up to 90 lb of chemical fertilizer per acre. These plants also concentrate a higher proportion of the annual growth in the seed head and have short stems so that the heavier seed crop will not 'lodge' or bend the stem down and make it impossible to reap by mechanical means. These new varieties are also quick growing and, given enough water and fertilizer, two or three crops may be sown and harvested in a year.

Leaving aside the immense technical, social and economic problems in supplying the required inputs and distributing the harvest, all of these new tropical varieties still have some basic biological limitation. After several generations the closely defined genetic characteristics of the population degenerate. At the same time, any genetic resistance to pathogens will be sacrificed along with the degeneration while the pathogens themselves will in time evolve new mutagens which will be able to attack any plant whose resistance persists. The farmer must therefore rely on a new supply of seeds each year and the plant breeder must continually develop new strains to remain one jump ahead of pests and diseases and to sustain the vigour of the new hybrids. The gains in crop production from genetic manipulation may be short-lived, especially if widespread cultivation of the new varieties leads to the loss of reserves of genetic variability present in the older varieties they replaced. To develop, test and plant a new variety also takes time—probably a minimum of 10 to 20 years. The rate of degeneration and of increasing susceptibility to disease is in many cases faster than this. So even the most enthusiastic promoters of new high-yield varieties admit that genetic manipulation cannot keep food production increasing faster than population for much more than two decades.

Pest and Weed Control

Besides an assured supply of inputs, modern agricultural systems require artificial controls over the attempts by nature to re-colonize and continue the succession towards a climax community. The controls must also prevent productivity being reduced drastically by rapid increases in the population of species which feed on the crops or pathogens which may destroy it.

The suppression of insect pests by DDT and other organochlorines, organophosphorus and heavy metal compounds has become widespread and residues of the more persistent compounds have been taken up by ecosystems resulting in death and sterility to non-target species, especially those at higher trophic levels and some sensitive aquatic invertebrates. Not only are the ecosystems affected but the relationship between the pest and its natural predators may also become disrupted. By killing the predator as well, future outbreaks of the pest may become even more severe or may be replaced by other pests. In Kenya, the cause of an outbreak of the Giant Coffee Looper on coffee plantations was traced to removal of its predator by parathion spraying directed at a target species different from either the looper or its predator. The looper was not even considered a pest of any importance before this outbreak [11].

Besides the side effects on ecosystems the pests themselves develop resistance to continued applications of pesticides. Since insect populations arc large and breeding rates rapid, mutagens with resistance (i.e. behavioural avoidance of the chemical or the ability to metabolize the insecticide) can proliferate fairly quickly. It is now reported that a total of nearly 250 species of insect pests are resistant to DDT [12]. The time interval between the first pesticide application and the development of resistance ranges from two to seven years. As a result new pesticides must be continually developed and the next group of pesticide compounds (e.g. organophosphorus, carbamates) is tried out as resistance to organochlorines is complete.

To overcome the problem of resistance, other forms of pest control are now being developed. These include breeding and releasing predator species, release of sterilized males to reduce pest breeding and use of hormonal attractants to trap and irradiate males. Another approach is to diversify the crops so that pests cannot expand to epidemic proportions. Obviously integrated pest control and multiple cropping bear a closer resemblance to natural conditions and in the long run control is more effective.

Hormonal herbicides such as 2-4-D act by selectively affecting growth rates of dicotyledon weeds while having much less effect on monocotyledon cereal crops. Other types may stimulate seed germination of weeds so that they sprout too early and are killed off by frost or other climatic conditions. Genetic resistance is also developing among weeds and there is some evidence that the use of herbicides may be reducing yields and rendering crop plants more susceptible to disease [13].

Limitations on Ocean Ecosystems

Nearly all these artificial controls and mineral inputs are subject to diminishing returns, have some sort of deleterious effects on natural ecosystems, especially aquatic ones, and consume substantial quantities of energy derived from fossil fuels. With these problems and the limitation on prospects for land-based food production it is hardly surprising that the myth of the vast productive capacity of the oceans is eagerly embraced by some as the panacea for closing the gap in food production.

Yet, as is only too well known by marine biologists, the maximum sustainable yield from the oceans is not much more than twice the present annual sea catch of about 52 million tons. Higher yields than this will require moving down the food chain from large fish to plankton but it is questionable that greater efficiency would result since more fuel and human energy would need to be spent on harvesting. Adding nutrient inputs is not a practical means of increasing productivity although proper management by protecting breeding grounds and improved cropping techniques may do so.

The 5-6% growth in total catch between 1950-1969 has been the result of technological improvements in tracing, catching and handling fish. Unfortunately the free-for-all situation with regard to harvesting has reduced some fish populations to dangerously low levels. The North Atlantic Salmon and Cod fisheries and many Tuna fisheries show signs of over-exploitation and of the Cetaceans the Blue Whale is estimated by some to be on the verge of extinction. The maintenance of even the present catches will not be possible for much more than a couple of decades if it continues to be based on traditional stocks. The large catches of fish for conversion into meal to feed poultry or to make fish protein concentrate is one way of utilizing species unacceptable for direct human consumption. Increased acceptance of squid species which are readily eaten in Japan and the Mediterranean area would also help. Where inland fisheries are concerned. exploitation, nutrient inputs and breeding rates are more controllable and present estimates of 40 million tons annual production could be increased considerably.

Conclusion

The process of agricultural development is exchanging diverse, stable food-producing systems based on renewable solar energy and small scale manipulation of natural rates of water and nutrient cycling for a system based on a small number of crop species with reduced genetic diversity, requiring sophisticated control mechanisms, high concentrations of inputs and substantial supplementary energy from limited reserves of fossil fuels. It is therefore fallacious to equate the present agricultural development process and the adoption of industrialized food production techniques with progress in terms of sustainable increases in yields and efficiency in conversion of resources to edible calories and protein.

The main objective of this process of agricultural development has become the assurance of food supplies for an expanding population. Since, in ecological terms, continual population expansion is not possible then perpetually increasing food production to catch up is as fruitless as chasing rainbows. At the moment it is also creating a dangerous illusion that ecological limitations have been overcome rather than just postponed. Sustained food production within these ecological and resource limitations obviously requires a stable or reduced world population. Unless this can be achieved, more and more of our energies and time will have to be devoted to food production rather than other pursuits. This will be the result of diminishing returns as well as population pressures.

With this as the future prospect one cannot help but be envious of the farmers and fishermen of the Lower Volta in Ghana whose traditional standard of living was described as an 'impediment to economic progress'. The problem seemed to be that with no shortage of land they could produce enough food by working a daily average of 4.7 hours for only 174 days in the year. They did not consider ostentatious consumption a virtue and free land did not mean that it had to be populated [14]. The most ecological and humanitarian approach to agricultural development would be to remove population expansion as the prime generator of the need for the adoption of industrialized food production techniques.

References

1. Gates, David M. 'The Flow of Energy in the Biosphere', *Scientific American*, Vol. 225 (3).
2. Rappaport, Roy A (1971). 'The Flow of Energy in an Agricultural Society', *Scientific American*, Vol. 225 (3).
3. Odum, Howard T (1971). *Environment, Power and Society.* Wiley-Interscience, p. 117.
4. Kleiber, M (1961). *The Fire of Life*, Wiley and Sons, New York.
5. FAO, *Small Farmlands can Yield More.*
6. Waller, R (1971). 'Prospects for British Agriculture.' In: *Can Britain Survive?*, ed. E Goldsmith, Sphere Books, London.
7. Blaxter, K L (1970). 'Comparison of Farming Systems.' In: *Factory Farming*, Symposium of the British Association for the Advancement of Science. Educational Services, London.
8. Ehrlich, A and P (1972). *Population, Resources, Environment—Issues in Human Ecology.* W H Freeman and Co., San Francisco.
9. *Unesco Courier*, July 1971.
10. *Observer*, 5th March, 1972.
11. Bigger, M (1969). 'Giant Looper, *Ascotis selenaria reciproca* Walk, in Tanzania', *East African Agriculture and Forestry J*, Vol. 35, No. 1.
12. Environment Staff Report (1970). 'Diminishing Returns on Pesticides', *Environment*, Vol. 12.
13. British Farmer and Stockbreeder, April 1971.
14. Lawson, Rowena M. 'Innovation and growth in traditional agriculture of the Lower Volta, Ghana', *Journal of Development Studies*, IV (1967). Also in: Wilkinson, R G (1973), *Poverty and Progress*, Methuen.

Discussion

MILES DANBY
I would like Mr Holliman to say a little more about his most
interesting comments about the farmers and fishermen of the
Lower Volta in Ghana, people who seem to have found a
remarkable equilibrium with their environment.

JONATHAN HOLLIMAN
My paper was primarily concerned with the ecological pressures
resulting from development programmes, but I included that
reference to serve as a reminder that social effects of
development are also of considerable importance. The people
cited in this example had a life style which appears, at least, to
be simple, but stable and satisfying. Their work load necessary
to sustain them is low, primarily because the area in which they
live is not subject to severe population pressures, resulting from
their cultural approach.

One question this raises is of the social aims of
development. Do we necessarily need all the encumbrances of
western life, with the ecological impact which they entail, to
live a satisfactory and satisfying life? This is usually an
assumption in setting development targets, just as it is an
assumption that because a new system is more profitable, it is
therefore more desirable, leaving aside more or less entirely the
ecological and social effects of the system in question.

SARAH WELLS
Going back to your reference to the Lower Volta area, isn't it
true that the fishermen have now moved up to the new Lake
Volta and are engaging in fishing which is more commercially
orientated?

JONATHAN HOLLIMAN
I was not trying to argue that the system under which they operated was necessarily the best one, but apart from possible social merit, it did appear to be a sustainable agricultural system. Their current life style may remain sustainable and involve a cash income from fish sales. This is no bad thing.

RANDALL BAKER
One of the systems which has been most disrupted by recent change is that of pastoralism, which has suffered wholesale disruption in many parts of the world, partly by the introduction of administrative barriers, partly by technological innovations and often by the actions of an unsympathetic government elite.

This is especially significant in view of the importance of pastoralism as a means of producing protein in arid and semi-arid areas. An extreme example of this disruption is evident in Saudi Arabia where, if present trends continue, there will be no desert Bedouin at all in twenty years' time, and if there are no Bedouin then there will be no production of meat from the desert and semi-desert areas. All this is taking place in a country spending £237 m per year in food imports for a population of 5 million, and annually using 870,000 animals for religious purposes, equivalent to about 29 million cans of corned beef!

JONATHAN HOLLIMAN
It is a fact that many people concerned with development do consider pastoralism to be a very backward way of life, even though it may be a system which can be sustained and is a highly effective way of utilizing marginal land.

OWEN JONES
Just to go back to the Lower Volta example, is it known how population stability was achieved in that area?

JONATHAN HOLLIMAN
I do not know the answer for that particular area, but it is often true that cultural controls do exist, working in various ways. These may not completely stabilize populations but do tend to prevent the huge surges which we have witnessed in the past 20

years. The recent phenomenon of population growth may be due, in part, to disruption of such culturally centred population control measures as well as to control of infectious diseases and other causes.

Development and Physical Resource Utilization

Anthony Tucker

Since I am a journalist I tend to react to things fairly quickly, and one of the things that has struck me so far about this conference is its sense of gloom. Maybe that is appropriate, but I would have thought that the element of optimism in most of us would tend to counteract gloom. Take for example this morning's delightful news that President Nixon has discovered that there may be an energy crisis. We have been telling him this for the best part of ten years. Now it's official! Behind the scenes presumably, in the United States, a great deal of work has gone on to find a way out. One thing that many of us feared was that as soon as the President discovered that there was possibly an energy crisis, there would be a permitted deterioration in environmental standards. Solving the energy crisis comes before everything else. This, of course, is already happening. The very stringent air quality legislation which is not yet even effective in the United States, seems likely to be thrust aside in the interests of using coal instead of natural gas. Such a hasty rejection of recently and hard-won standards shows us that energy is really one of the most important of resources, and I am going to talk about it a little bit more later on.

I remember, about 1966, I was wandering around Tromaby, near Bombay, the Indian Atomic Energy Commission Research Centre, with Homi J Bhabha. He was a very interesting man, head of the atomic research and head of many other things too. He met an unfortunate end on a mountain in Switzerland at the hand of aerospace technology. But in 1966 we were looking at the way the Tromaby laboratories were being built. It was, to me, a very strange situation, because a lot of brick-lifting was being done by women, who were climbing up crude ladders made out of lashed poles. Yet, inside the crudely built laboratories was a very high technology indeed. Homi Bhabha turned to me, smiled, and said: 'You know, we can never get a door to fit—but the developments going on here are the most important in India at the present time'. There was an absolute dichotomy; on one hand extremely primitive building techniques and on the other the successful pursuit of the most advanced of technologies.

Among recent arguments which have arisen about the ways of inducing a rapid achievement of development potential within the less developed countries, or in many of them, is whether or not to impose high technologies. This was an argument put forward by Herman Kahn some time ago and repeated here by Angus Hone. It is said that there can be a transfer of technology from, say, a multi-national company-type of organization, because this type of organization is extremely interested in low cost operation and in particular in low cost labour which it can find in a developing country.

This is almost exactly the Victorian situation in England. Manchester 1832 on a grand scale. The question is whether this is even a permissible way of letting things begin. It assumes, first of all, that there would be a transfer of technology. I don't believe any of the multinational corporations, or indeed any of the advanced technology organizations in the western world *transfer* technology; they sell it; they use it as a means of barter; they will use it to distort or exploit a situation so that they can get some kind of pull-back from it: but they will not transfer it.

If, for example, India had waited for somebody to come along and present them with nuclear technology, they would not have got very far. They had to put immense effort into it themselves; they had to have men of the right calibre who went out and gathered expertise. The starting point was in India itself, not outside it.

Now the starting point for a multinational company is not in India or in Ghana or anywhere like that—it is inside a multinational company whose prosperity and self-interest is its own greatest concern. To be sure, a company's investment in a developing country can lead to growth, but only economists confuse development and economic growth in the sense of implying that they are the same thing. Indeed the two terms are often used interchangeably by economists—they are, of course, entirely different things. It may be that to achieve some kinds of development you must have economic growth; maybe for all kinds of development you must have economic growth of some kind, but economic growth itself has no meaning or quality whatever. It is just a measure that has nothing whatever to do with the real nature of development.

Development is something which effects the culture, the fabric of a society, the way it works, its desires, values and goals. It is about the way individuals express themselves, whether they express themselves fully and whether they can express themselves more or differently. If they move in a direction that is socially desired then that is development.

It follows that the kind of thing that the western economist and the western industrialist thinks of as development may, in fact, not be development at all. It may be regression. I just want to quote you one or two people on this. You must know the publication, 'Development Forum'. Judge Jaggon (Ghana) writing on this kind of subject says: 'Efforts to get commodity price agreements have been very firmly resisted by the industrialized countries. It took 16 years to reach an agreement on the minimum price of cocoa, and after those 16 years, immediately after the agreement was reached, the richest countries in the world made reservations so that the agreements were no longer applicable'. That is the way the difficulties of the underdeveloped countries evolve—in a sort of vicious circle. It is as if the developed world is unable to understand the process and the effects of their own economy and behaviour and the way these oppress the developing world. Judge Jaggon goes on to say that now that 'the developing countries need the markets of the developed world, and immediately the developed world finds the developing countries going into manufactures which could close the gap, it sets up barriers, discriminative barriers, against trade'. That, I suspect, is a view of the common market from Ghana, and it's a fair one.

Here is another point of view: from Samir Amin. He is the Director of the African Institute for Economic Development and Planning and he says that 'the choice of growth instead of development is not an accident'. He's talking about the way the word growth, economic growth is used instead of development. 'The choice is not an accident for it helps perpetuate an imbalance that has served the capitalist world well. The imbalance concerns the International Division of Labour under which the industrialized nations of the centre, that is the developed world, concentrate on high technology and a high return on manufactured products while the developing countries have the periphery—concentrate on raw material production and light manufacture. The aid policy of the West, focused as it is on a few major objectives, tries to ensure that this division of labour continues for ever.' The objectives, says Amin, are the stabilization of raw material prices (we're back to cocoa now) with buffer stops to narrow price fluctuation, and the promotion of a few import substitution industries, which are the industries that the multinational corporations would like to set up, plus the attainment of certain quantitative aid goals. The encouragement of foreign investment, particularly that of the multinational corporations, is all designed to achieve economic growth. But it is growth that does not change the pattern. It leaves the division of labour intact and keeps the economies of the periphery on the same well trodden, low level tracks.

These are views that are not frequently expressed over here, although they may be among yourselves, but it seems to me that the efforts of the West to help are seen from that side as a massive technological trap, as though we are always doing the wrong thing, aimed at perpetuating a situation that we ought, in fact, to be changing. It is not that we are simply leaching resources from developing countries, who are very soon going to need these resources themselves, but we're actually perpetuating the gross imbalance and the impossibility of true development in those areas into which aid, as we call it, is being sent.

I often wonder whether it was an accident, or whether it's very significant that the 'Green Revolution' is conceived as a kind of crash programme to provide food or whether it is to expiate a Western guilt complex. For the West initiated massive population problems by technological accident and failed to think out, properly, the routes to improve agriculture in developing countries. This really requires a very careful examination of the agricultural practices and conditions in various parts of the world. It would not then have been based on remote efforts of research and an attempt to impose ready-made techniques. Anybody who is involved in farming or in gardening knows that in any one place you have problems that you don't have only a quarter of a mile away, and that you have to adopt great flexibility in the techniques you use, even where conditions appear to be similar. If this was not evident to the people involved in the thinking behind the 'Green Revolution' then they knew nothing about the problems of food production, and should not have been doing the research.

There is now a transition, of course, with a serious attempt being made to find out what the conditions are in specific places, what the ecological problems are, to actually develop agricultures that will work, but I do not know whether anybody is thinking, in a long-term flexible way, of maintaining a continuous transfer of knowledge to those centres in the developing world where such knowledge can be applied at the grass roots level.

One major goal must be to overcome the limiting effects of market value, where the man with land has the grip on everything and can make the decisions which keep production at the point where his return is best, or where the smuggling prices are highest.

I'm not suggesting that you can cure overnight the kind of social malaise that exists in many developing countries and which promotes the tendency to corruption. The men often technically in charge are well down the social spectrum and may be corruptible. Yet if effort could be fed in to change subsistence farming by *practical* means, not by having to buy things but by better techniques, there could be a general upgrading of the standards from the lowest end. This may be a different approach but, instead of high technologies, it may be that we should be thinking about it in that way.

But to go back to energy. Figure 1 shows recent American and European predictions for energy demand in various countries throughout the world. The growth rate in a number of situations is very much faster than in the United States. The United States is still exponential but there is a closing of the gap, evident here, in real terms. Of course the amount of energy growth in the United States is far greater than any other country, but over this time span the developing countries together comprise a rate of growth that is very much faster than the United States itself. If this turned out to be true it would be very important.

Figure 2 is another view of the prediction. Africa, and Western Asia are, roughly speaking, at the same energy demand

Figure 1.

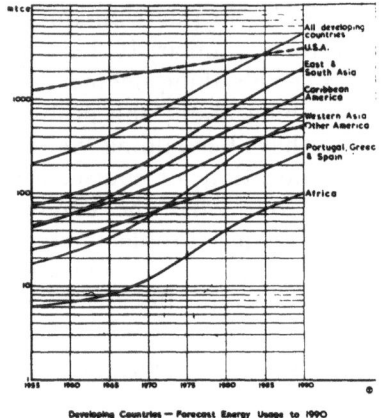

Developing Countries — Forecast Energy Usage to 1990

Figure 2.

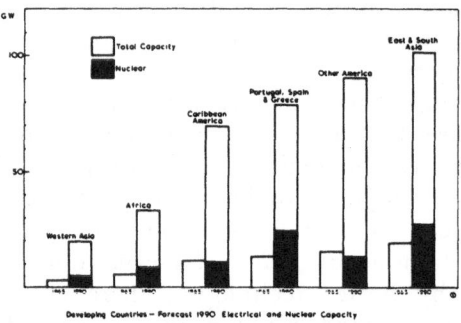

Developing Countries — Forecast 1990 Electrical and Nuclear Capacity

level as Britain, but clearly the gap between them and the developed countries would grow if this prediction is right. One of the interesting things about these predictions is that they cover the decades during which the medium term energy crisis is going to strike. The growth rate of energy demand in the developing countries which may exceed that of the United States before 1990, is accelerating at a time of shortage. We have the United States, Europe, Japan and now the developing countries, all requiring roughly the same amount of energy by 1990. That is between six and ten times present world consumption. Where is it going to come from? United States' geologists are scattering about in the fringes of the arctic, right up in Northern Canada, trying to find oil. We know there is coal in Bangladesh, very deep, but the major sources of oil will for ever be the Middle East and the various shales and tar sands in the United States. Is the United States going to sell its oil to the developing countries? I don't think so. Will the OPEC countries—a group which has the Western world very tightly in its grasp make the kind of two-level price system which has been suggested here? Or will it utilize this very valuable resource at maximum profit, in imitation of Western practice?

I think that OPEC is going to hold on to its resources and not believe, as many of the Western technical experts are trying to persuade us, that nuclear power will replace oil within 30 years. If OPEC takes that kind of stand it can control the developed countries and help the underdeveloped because it controls sources of energy. Further, if this kind of OPEC system works, the possessors of other resources may get together and say, 'these are *our* resources, we want to market these in *our* way'. It seems conceivable that other raw material, which at the moment the West is importing at relatively low cost, such as copper, chrome, nickel and almost all the basic metallic ingredients for high grade alloys and so on, could be properly controlled at source. Revenue could be raised at very much higher levels than at the moment from DC to UDC.

The DCs are, in fact, highly vulnerable. At the moment, for example, the US is breaking UN sanctions in order to get chrome from Rhodesia, only because it can't get it from somewhere else. That is a very important fact which the UDC's can use, politically and commercially. Yet even assuming that such pressures begin to transfer funds from the developed to the developing countries, there's no guarantee, that having arrived there, the funds will achieve anything, for, as history tells us, development has to be based on a strong agriculture if it is to succeed. But the agriculture of the world is weak, and it is getting weaker. Its requirement is a greater energy input, and we have already seen that we are already running into energy problems. There may be miraculous answers, in the form of fusion power or something else, in 30 years' time but long before then we are going to need a much greater input into agriculture in developing countries. New, cheap, yet available techniques are desperately needed.

The French, at the moment, are playing about very

successfully on the edge of the Sahara with a quite simple solar still, whose purpose is to distill brackish or salt water for agriculture in an arid region. It has been found that a very simple system, costing probably less than a thousand pounds, can transform about a quarter of a square mile of barren wasteland into a garden. Has this been tried in India? I don't believe it has, either there or in other drought ridden areas. Why hasn't it been tried? We could have built cheap stills quite successfully in 1920 if we'd wanted to do so for there's no new technology involved. But it didn't happen in India.

The reason, I suspect, is that solar stills embrace technology alien to the energy attitudes of the Western world, yet the developing countries have, in general, the benefit of a massive solar flux. Let me give you some figures. At the equator the solar input is about 2,200 kilowatt hours a year per square metre, which is roughly the energy output of a pair of bullocks. That, I would have said, is quite useful. Within the tropics, the figure is about 1,900 kWh, at mid-latitudes about 1,300-1,500, and even in the middle of the UK is 1,400. This is a surprising amount of energy.

It is rather interesting that in attacking the Western approach to aid, and listing some of the technologies that the West keeps to itself, a Mr Amin includes utilization of solar energy. I do not think there is any high technology needed, but there is certainly a great deal of developmental skill required. But it is the kind of skill which shows how to build systems at village level, permitting the fabrication of light goods at a village level, and beginning what I would call the upward industrial spiral.

The trouble with solar energy is that it is dilute. You can't build a 2,000 megawatt power station as we build them for coal or oil, but you can quite easily collect enough energy to drive light industry at a village level. That seems to me to be an important fact that nobody has really bothered about. Certainly the technology of the West has never looked at it seriously, although in the United States, because they are facing a crisis, a great deal of new money is going into investigating how to make solar energy do the work that, at the moment, the electricity system is doing against the input of solar energy. You have a situation in the US in which peak demand comes in the summer when everybody's pumping heat out of their houses with the air conditioning system while pouring electricity in. Why not use

the solar input to do this? I think it poor engineering to do it
the other way, and there are more possibilities there than have
been looked at seriously.

It is especially interesting that the highest inputs of solar
energy are largely in those areas in greatest need of energy. If all
other energy sources cease, we, in the developed areas of North
America and Europe would find ourselves very much lower
down the energy scale than we are at the moment. In fact, when
you look at the distribution of genuinely basic resources like
solar input, the availability of metal ores, the greatest potential
plant growth rate, and even the availability, for example, of a
temperature difference between the surface and 300 feet down
in the sea which could be used for energy production, it all
seems to lie between the Tropics. That may look good but, my
goodness, we need to get at it quickly. India's crisis could be
terrible within a decade. Going back to the brief time I spent
there I was very surprised that there seemed to be a genuine
feeling among people such as engineers, that the situation was
already impossible. We have got to throw three or four
generations away, because we can't do anything about it, they
would say. I came back from India very sad and very sick
because I believe that we could do something. I think the West
could do a lot about it but won't because its thinking and
efforts are wholly strained by considerations of economic
growth.

This is the absolute crux of the problem. I am not suggesting
that we do not need economic growth, but that this is not
always sympathetic with the real aims of development. You
know, if the solar still system can be made to work, if one bit of
investment at that sort of level can turn a quarter of a square
mile of arid countryside into something very much like a rich,
intensive farm garden, then goodness me, it's worth looking at
even if our commercial masters cannot see a quick buck in it.
Agriculture is the key to real development. The key from the
developing country's side may be that they have, in many ways,
the resources that we want, and they could manipulate us if
they use that knowledge. But we have knowledge and ingenuity
which, at the moment, we are *choosing* not to transmit. This is
to invite hostile manipulation.

Discussion

G S PURI
What are your own views on the pattern of development you mentioned where high technology goes on side by side with primitive methods, as for example in the case you quoted from India?

ANTHONY TUCKER
In that particular case, the development of nuclear power was logical after the decision had been taken to develop an electricity grid on the basis of hydro-electric power from Northern India, and the amount of investment involved was small compared with many other countries. India had plentiful supplies of thorium and needed cheap energy sources to power industrial developments. The question is really one of whether the original development of the grid was desirable compared with other uses of the funds available, but once that decision had been made then the subsequent decision to go in for nuclear power is logical.

This touches on an important point, namely the way in which an initial decision in a development strategy will influence a whole range of subsequent decisions, setting a pattern for a long time ahead.

RANDALL BAKER
I would like your comments on two points, both related to items you have mentioned.

The first point is that one of the most important single problems in international development is the problem of balkanization, the existence of inumerable small states, none of them viable in economic terms and most trying hard to industrialize although such processes cannot be efficient because of market size. Even joint bargaining by groups of states on commodity issues rarely mean very much. Even the OPEC negotiations have resulted in increases of only a fraction of a penny per gallon in the price paid to member states for their crude oil. Saudi Arabia now gets about 0.75p. per gallon which hardly compares with the price we pay for the petrol as consumers. The coffee producers of Uganda get 3p per pound for their coffee; we pay a retail price of 100p. If we get things into perspective, then the chances for success in bargaining by groups of small states are surely minimal.

My second point relates to the need for cooperation in research. We have talked about the Green Revolution on several occasions here and I think one must be careful not to run down the remarkable technical achievements which have been involved. But the problems which have arisen are examples of two faults in research practice. One is the danger of transferring attitudes along with research results and the other is the lack of interdisciplinary contact. I am convinced that the greatest need is for multidisciplinary research programmes. Such programmes might have pointed to the problems of enhancement of social inequalities incumbent in the Green Revolution.

ANTHONY TUCKER
Taking your first point, I would agree entirely that present units are far too small. It is largely a matter of increasing political sophistication and commercial understanding, but I think that attitudes have changed remarkably quickly in the past five years and there is plenty of hope for greatly increased cooperation among small states in the future.

Concerning research, I did not intend to disparage the work that has gone into producing high yielding plant varieties but I do believe that much of the philosophy behind it was wrong with reference to the target at which the research was aimed. The kind of technical aid that might have been more useful would have included, for example, on-the-spot technical assessment and evaluation of existing subsistence techniques and methods of improving them, evaluation of varieties already available, detailed water requirement studies and ecological studies of existing host-parasite interrelationships.

All these could have helped to improve the agricultural system concerned without inducing social disruption in the process. For research inputs to be effective they must be related to the actual situation in the areas concerned as well as national and international political considerations. I really don't know how one achieves this, but I think that a lot of current technical assistance is misdirected. Far more attention should be paid to the local situation, especially to the matter of training people at that level.

Does anyone seriously think that we are going to meet the demands of population growth by the turn of the century on present form? I think it is impossible.

KATHRYN MORTON
You say we will not be able to feed the population that is to come. Isn't that unnecessarily pessimistic? After all, if we had known in 1945 that the population would rise to 4 billion by 1975, we would have thrown up our hands in despair.

ANTHONY TUCKER
I am not throwing up my hands in despair. On the contrary I think we have to be realistic about the situation if we are going to make the changes necessary to cope with it.

DAVID HALL
I would echo what you have been saying concerning development but would make the point that much of the UK aid effort has been directed at the grass roots level. This is particularly true in the area of education and also in agriculture where there is a great deal of effort in the field of extension, especially trying to understand the ways in which subsistence farmers operate. It may then be possible to stimulate the sometimes minor alterations which can lead to development. As you rightly say, development is not necessarily equated with economic growth.

On another point just discussed, I agree about the disadvantages of small states and would make the point that although we may consider regional groupings advisable, this is still not the general view among the countries concerned.

The Implications of Ecological Limits to Development in Terms of Expectations and Aspirations in Developed and Less Developed Countries

Edwin Brooks

Human history is littered with prophecies of doom, and the more portentous soothsayers find a receptive audience. Thus Malthus gained masochistic immortality in 1798, as did Marx and Engels half a century later. Today, of course the entrails can be more precisely quantified, but it is doubtful if the modern sense of an impending ecological catastrophe is ultimately due to system dynamics and computer models. More likely it springs from the ancient sense of lost innocence which Genesis related to expulsion from the (godly) natural order, plus the alienation of man from man which Marx related to class deformations of human community. Expelled from the Garden, carrying the mark of Cain, and valued as commodities on the labour market, it is small wonder that we behave in an anguished and somewhat schizoid way. We are simultaneously predators and social animals, forming friendships in order to combat our enemies. We assume we are the lords of creation in unfettered command of the planet, yet throughout history we have desperately tried to placate the huge forces of a hostile nature. We have developed, through speech, a unique ability to co-operate with other individuals of the species, yet even our linguistically homogeneous nation states show that we can be just as ruthless towards one another as towards the brute order and our habitat in general.

This may seem a far digression from the theme of this symposium, yet I suggest that the various reactions to the Club of Rome's notorious report. 'The Limits to Growth' can only be properly understood against such a background. For example, those prone to believe in Fallen Man whose sinful pride has debauched God's purpose in history, are not ill-disposed to computers verifying their late-millennial forebodings about the Day of Judgement. Those who believe that we are born to demonstrate our moral fibre in the face of tribulation, nod gravely when the world model demonstrates that all our technological precocity cannot bring a consumers' paradise upon Earth, and that terrestrial utopia is for ever beyond the reach of the teeming multitude of men.

Conversely, those who believe in the Original Virtue of the proletariat, or the wretched peasantry, will dismiss ecological warnings as the latest addition to the bulky pharmacopoeia of opiates for the masses. Just as Marx decried Malthus as the apologist of those who wished to keep the poor in their place (i.e. pressed against the subsistence line), so the Club of Rome is denounced by some as the authors of a neo-colonialist plot in which the poor (and coloured) peoples of the world forget their ambitions of affluence. Either way, the statistics presented by Professor Dennis Meadows and his colleagues tend to be viewed through the clouded prism of political and religious doctrine, and in such circumstances it is hardly surprising that the persuasiveness of the Limits of Growth thesis (or what is supposed to be its thesis) will vary both within and between the countries of a deeply divided world.

Yet it would be too simple to dismiss this welter of contradictory reactions as mere proof of the inherent subjectivity of 'facts'. For lurking behind the suspicion that eco-catastrophe is a subterfuge of the rich to maintain their dominant industrial status, is the undoubted fact that the privileged position of the wealthy countries is bound to be undermined once the poor begin to claim a fairer share of the Earth's resources. Moreover, as soon as the demand for a larger slice of the cake can no longer be assuaged by promises of making the cake itself vastly bigger—and if the ecological constraints are valid then this hope is bound to prove a mirage—it becomes clear that the more the poor can be persuaded to curtail their resource demands, the more will be available for the booming economies of the developed world.

Thus the first, and in many ways the central argument I wish to advance, is that the greater the ecological constraints upon growth, the more we shall find the whole environmental argument focusing upon the global distribution of wealth. Moreover, since the allocation of this wealth is fundamentally a reflection of the power configuration of the world, the tightening of ecological constraints is bound to produce tumult and tension in international relations. In short, the recent battle over oil prices is merely the first major engagement in a protracted and potentially violent confrontation between the Have and the Have-Not nations. The outcome of this struggle is clearly impossible to foresee, but it seems distinctly premature to anticipate an egalitarian solution. If indeed we are faced—on the planetary scale—with the economics of scarcity rather than abundance, then the politics of conflict seem more probable than those of co-operation and friendship. In such an arena, it by no means follows that the poor country will be the victor; the success of OPEC during the early seventies may have been due to peculiarly favourable circumstances, in which the bargaining power of the Have-Nots was aided by a geographical concentration and (general) unanimity of purpose between the major producing countries.

There is an obvious analogy in fact between the problems of working class organization in Victorian England, where without unity there was certainly no strength, and the present task of the Third World in negotiating from strength with the advanced capitalist and communist countries. But whereas in nineteenth century England the economy was able to grow without ecological limitations (on the contrary, Malthus's 'finite' Land was effectively enlarged immeasurably by the opening up of new areas overseas which became tributary to Britain), no such empty and fertile areas remain to be discovered by the Third World. Thus the renowned ability of the British ruling class to throw imperialist crumbs to the workers, and thereby diminish their militancy and revolutionary spirit, is denied to the governments of the developing countries. But equally, to the extent that ecological constraints press upon the world as a whole, the governments of the developed countries will find it difficult to find surplus wealth—other than by taxing their citizens—which they can hopefully send to dampen militancy in the Third World. They are most certainly not likely to find the political will to do this until such time as their citizens come to recognize that their security and their very survival depend upon such a transfer of wealth.

During the First Development Decade of the 1960s, the argument in the developed world about assistance programmes proceeded in a rather rarified way. Notional targets were ostensibly set, which even had they been met would have been very modest, and in any case as Mr Dinwiddy has made clear, they were hardly ever reached. Occasionally, as at the various UNCTAD sessions, the cry was heard for an end to wildly fluctuating commodity prices, and for ensuring instead that stable guaranteed prices be paid for the Third World's produce. But here too very little was accomplished, and against a general background of deteriorating terms of trade against primary producers, the seismograph of price graphs was only too clearly matched by political instability in the primary producing nations. The lesson became painfully obvious: that world trade was not a relationship between equals haggling, as it were, in a free market place. Instead, as Professor Dahlberg has shown, there is 'a natural tendency for resources and information to be extracted by the richer, more organized subsystems from the poorer, less organized ones'. And the 1960s showed that this natural tendency, whereby the rich get richer at the expense of the poorer, was sufficient to over-ride the frail effort via assistance programmes to redress the imbalance.

The question which is now posed to the Third World is implicit in Professor Dahlberg's own plea that 'international development programs ... should be conceived in terms of counteracting and placing limits on (this) natural tendency'. How, in short, is this pious hope to be realized? The OPEC saga, which is far from having run its course yet, suggests that the answer will only be found via coercive measures applied by the producers of increasingly scarce and therefore increasingly expensive commodities. To the extent that they can negotiate from strength, and thereby hoist these prices still higher, they will be doing no more than has already been successfully accomplished by organized labour in the liberal democracies. It is doubtful, alas, whether organized labour in those rich countries will regard such successes as victories in a common struggle. More probably, they will be seen as undermining the recently won affluence of the industrial proletariat in the developed countries, and as with the rapidly rising price of petrol there is bound to be a conflict of interest between the rich conspicuous consumers and the poor conspicuously non-consuming masses of the Third World. Out of such a potential and long-term conflict of interest, with different races involved in global confrontation, could easily rise new tyrannies and new racist ideologies calculated to maintain the privileged status of those with so much to lose from a policy of equal shares.

For this reason, as I suggested earlier, we cannot automatically assume that the world society which is going to have to come to terms with its finite and crowded environment in the coming decades, will necessarily be egalitarian or particularly nice. Thus the second major argument I would advance is that the greater the ecological constraints upon growth, the more starkly will be at risk the liberal values of democracy, equality before the law, racial tolerance and so on. A revolution of rising frustrations and permanent confrontation is hardly the climate in which such fragile plants as liberty, equality and fraternity can flourish. In other words, if the extrapolations of the Meadows world model are confirmed in the event, then freedom will have proved no more than an ephemeral episode in the history of the human species.

This may seem to be stating the obvious, yet when we glance at the political priorities in the developed world it is hard to find much evidence that the implications of this situation have been even dimly recognized. The assistance programmes are generally trailing sadly behind the far from ambitious targets of the Pearson Commission, and the trading policies of the enlarged EEC give little indication of any horizons beyond the common external tariff. Even more ominously, the political parties in the rich states are still pursuing economic growth targets which are becoming manifestly absurd in the light of environmental limits. Here and there we see signs of a re-appraisal—as with the recent British awareness of a looming energy crisis in which the long despised coal measures are going to have to play a much more important part than was assumed throughout the 1960s. Even more belatedly, we are coming to recognize that the age of the petrol-driven motor car is going to survive about as long as did the Age of Canals in industrial Britain, and that even the North Sea discoveries will not prevent the mounting shortages of oil—nor its soaring price—which will dominate the industrial world economy of the 1980s.

Yet so far the logic of facts has barely been perceived by the decision-makers, and given the compulsions of democratic politics it is understandable that Micawber is preferred to

Cassandra. Technomanic ventures such as Concorde, which will surely be seen by future historians as the apotheosis of economic growthmanship (you can't stop progress), still forge relentlessly on, totally incapable of ever making a profit, certain to degrade the environment, and absurdly profligate in the burning of scarce fuels. Yet if after ten years of informed and justified criticism, and in the aftermath of cancelled orders everywhere, Concorde continues to win political favour, it is difficult to be optimistic about the rich world's ability to re-focus its political priorities.

The serious implications of this dilatory and myopic attitude by the rich countries towards conservation of scarce resources become all the more evident when we consider the population explosion in the poor world. Given the demographic structure of these poor countries, in which population pyramids rest upon widely spreadeagled bases of soon-to-be fertile youngsters, I would endorse Professor Newbould's belief that 'however successful population policies are, the world population is likely to treble before it reaches stability'. The global restraints upon development are therefore bound to become much more explicit as the planet finds itself carrying upwards of ten billion people, and as Professor Newbould goes on to argue this 'will require a reduction in the per caput resource use and environmental abuse of the developed nations to accompany the increased resource use of the developing nations, a levelling down as well as up. This conflict cannot be avoided'.

I would certainly not dissent from this forecast of conflict ahead, but I wonder if we have begun to grasp the magnitude of the political strains that will arise in the rich world as the 'levelling down' is forced upon it. The political and economic elites in 19th century Britain, for example, were to see their privileges and their dominant status challenged, and as with the extension of the franchise they had to adjust to a measure of power-sharing with the new proletariat. But they had at least the consolation of knowing that their absolute standard of living was not diminished, and that on the contrary their country's economic growth could simultaneously permit growing affluence for the masses and growing absolute affluence for themselves. There may be some exceptions to this broad generalization, as among the owners of stately homes who were slow to grasp the flexibility of our inheritance taxes, but there can be no doubt that the absolute standards of living of the better-off sections of society have risen substantially despite the concurrent political emancipation of the rest of the population.

The same consolation will not be available for the better-off sections of the world community—unless, that is, we argue that ecological constraints upon economic growth are a figment of our masochistic imagination. For those less euphoric, it increasingly becomes evident that our species is now proliferating (and for at least a generation to come is bound to proliferate) at a rate which is incompatible with the universal diffusion of those exceptionally high levels of material affluence which have recently been enjoyed by a privileged minority.

Against this background, three alternative scenarios can be postulated. First, that the rich world will be stripped of its privileges and wealth by the bargaining strength of the developing countries. This was the coercive process exemplified during the OPEC negotiations; but as I suggested earlier, there can be no certainty that such favourable circumstances would generally operate to the benefit of other Third World producers. However, coercion can take on many guises, and with nuclear weapons already spreading throughout Asia we cannot rule out the possibility of local conventional wars against the rich world, and its overseas investments, in which the latter's nuclear armoury could not in practice be used to offset the weight of numbers. The lesson of Viet Nam is that so long as the Third World country has powerful nuclear partners in the wings, it can take on and defeat a rich super-power. With China, and perhaps India too, likely to possess a massive nuclear stockpile and ample ICBMs by the 1980s, the scope for limited wars under the nuclear umbrella is unlikely to diminish as global tensions, envy and frustration accumulate.

The second scenario also postulates coercion and intermittent military engagements, but this time the picture is reversed, with the rich countries using their power nakedly to retain their privileges. There is a hint of some such trend in the tacit understanding between the two major super-powers about their respective spheres of influence; thus the Red Army is 'permitted' to invade Czechoslovakia with tanks, while the USA is 'permitted' to bomb North Viet Nam with napalm. Admittedly such examples of *real politik* (to which we might add in the post-1945 era Guatemala, Hungary, Cuba and the

Dominican Republic) serve to show how precarious are the conventions underlying the balance of terror, yet the common interest of the nuclear powers in seeing that their atomic weapons are never used might well develop into a tacit agreement—which even China might come to endorse—that troublesome and shrill countries in the Third World should not have delusions of affluence, run amok, and upset the applecart of global peace and security.

The third hypothetical scenario is presumably the one which the civilized liberals among us would prefer. That is, a world in which the rich countries accept that in their own long-term interests (including those of peace and security) it is vital to avoid a crowded glowering planet of massive inequalities of wealth buttressed by stark force and endlessly threatened by desperate men in the global ghettoes of the under-privileged.

In this alternative scenario, the basic strategy would have four main objectives. First, a massive re-appraisal of economic growth targets and procedures in the developed world, with maximum substitution of resource-consuming activities by resource-conserving ones. Second, an equally massive effort by the rich world, as well as by the poor world, to encourage economic growth as rapidly as possible in the latter. If this seems to contradict the first objective, let us be clear that the sort of economic growth being advocated in the Third World should not simply reproduce the classic Western model of speedy and sophisticated industrialization. Here I would endorse much of the arguments advanced at this symposium for encouraging agriculture (and peasant agriculture, too), and intermediate technologies; but I think we must also be clear that without economic progress in improving living standards and reducing the chronic malnutrition of the Third World, we shall fail to achieve the sort of motivation among parents which helped the demographic transition to lower rates of population increase in late 19th century Britain. The third objective follows on directly from this: there must be far greater emphasis internationally given to birth control facilities, and as an immediate priority an expanded scientific research programme to discover suitable safe techniques of family planning by the world's peasants. World Population Year in 1974 could be a watershed in this respect, and we can now hope to see the United Nations and its specialized agencies at last mobilizing in a manner befitting the scale of the task ahead.

Finally, and in one sense pervading all three of the other objectives, there is the need to define a new politics and a new morality of economic and social behaviour. The mainspring of political action is ultimately the Idea, or what the anthropologist might term the Myth. In earlier stages of political evolution the 'State-Idea' arose as the embodiment of people's perception of 'their own' community. The force of the territorial imperative which this Idea enshrined, as compared to the force of that other major ingredient in human behaviour, the sexual imperative, was neatly expressed by R. Ardrey when he reminded us of how many men have died for their country, and how few for love of a woman. The sons of London's slums died on Flanders fields for the sake of an England of green pastures and lowing cattle, village inns and quiet lakes, an England which none of them knew except with the eye of imagination sometimes known as patriotism.

Those who thrill to the pictures of Earth in space, in all its beauty against the infinite blackness, are perhaps beginning to grasp the new territorial imperative which is becoming necessary if the human species is to fulfil the promise implicit in its self-styled title of homo sapiens. The difference, of course, is that in the past the State-Idea arose to separate the insiders from the outsiders, and the nation-tribes gained a sense of community paradoxically by identifying and signalling their difference from the rest of mankind. In the modern world, as I have tried to argue in this paper, the old tribal divisions are far from dead, or even necessarily dying, and unless we can use the ecological argument to transcend the ancient group loyalties we shall find the ecological crisis inevitably sharpening the group conflicts. Ultimately, therefore, it would appear that the essential political implication of the looming ecological constraints is the clear choice we possess, between fashioning a new ideology of human brotherhood—with all its egalitarian consequences—or reverting to a world of hostile tribes desperately trying to procure and safeguard their shares of a dwindling stock of global resources.

So I would end, perhaps rather surprisingly in these days of system dynamics and electronic gadgetry, by suggesting that the ecological crisis throws into the sharpest possible relief the moral predicament of a species which gained mastery of the planet by its predatory skills, and which stands to lose that mastery by its difficulty in outgrowing that predatory ancestry.

Addendum

My original position at the end of the programme had led me to prepare an epilogue, and I therefore wore my parson's hat when drafting the circulated paper. But having now been shunted forward to a less godly time in the morning I shall revert to type and put on my politician's hat forthwith. This switch is not entirely unjustified, for the basic theme of my original paper was the changing configuaration and distribution of power in the foreseeable future.

It has struck me during this conference that overt conflict, that is war, seems to be visualized as a thing of the past. Indeed, I do not think that little three-letter word had previously been mentioned. Nevertheless, the briefest glance at the modern world shows that conflict remains everywhere latent, and is frequently explicit and violent. The clash of political ideology, of religion, of race, and of economic interest, is everywhere only too apparent. Deriving from this fact, my basic thesis is that growing shortages, dashed expectations, and frustrated demands will provoke ever more dangerous confrontations within a world which is already polarized socially, culturally, ethnically and politically.

I can illustrate an extreme and even exotic example of this clash of cultures by reference to some photographs taken during a recent expedition to the Amazon basin in Brazil. Here we can vividly see how the developed world of Rio, Copacabana, Sao Paulo or Brasilia is rapidly pushing into the once remote rain forest. There it inevitably comes into conflict with the aboriginal Indian tribes of the interior, who can overnight find their position threatened by the arrival of so-called *civilizados* who have little understanding or sympathy with their allegedly primitive practices and rituals. Last year I found myself meeting Indians who had been contacted by the outside world only months before my own visit to their jungle clearings. Their innocence of our ways was both startling and touching; I recall the naked lady who implored me to give her my socks, whose colour she greatly admired, and I shall long remember her proudly wearing them on her arms!

Such Indians are now finding the highly mechanized road builders lunging into their forest sanctuary. Thousands of miles of red earth highways are rapidly pushing back the frontiers of the inaccessible, just as did the railroads of the Old West a hundred years ago. From the air we can already see the fires which mark the sites of new towns in this scarcely mapped wilderness, and aerial photographs taken in the state of Acre show that quite extensive areas of the forest have already been burned off and converted to agricultural use. Perhaps all this promise of riches will prove a mirage, with El Dorado turning into a sterile and laterized desert, but for the present I merely cite Brazil's Amazonian adventure for the example it provides of my theme of culture-clash in a world of mounting environmental pressures.

The Amazon highlights the problem of two worlds—the developed and the underdeveloped—living side by side on the same planet (indeed, in this case the same national territory). It begs the question of how far we can insulate the weaker culture from the stronger one, and whether indeed we have any right or obligation to do so. It encourages us to ponder the motivation and morality of the richer culture; why, for example, it should bother itself at all with a mere handful of people, 150,000 Indians or 0.15% of Brazil's 97 million, at a time when so many others hope to benefit from the rapid exploitation of the country's interior.

In short, I feel that the fate of the technologically backward Indian is much more than an anthropological curiosity, and that instead it hints at our wider world dilemmas. Thus the official Indian parks and reserves which cover quite extensive areas of Amazonia, and in which the tribes purportedly have protected status, may have some bearing on the small 'impossible' countries of the Third World whose problems we have earlier discussed. How far can these helpless countries be given similar protected status, and how far can their infant economics be insulated from the ruthlessness of the competitive market economy? Is such paternalism desirable, and is it a practical objective in any case? Perhaps, as I suspect in Brazil, benevolence towards the least fortunate is already in short supply among those only little more fortunate, and a world of increasingly scarce resources will have even less sentiment to spare for naked savages and other wretches of the earth. In Brazil, where not only history is being telescoped, but pre-history too, the momentum of economic growth is unlikely to be slowed down by the special claims of those who need generations to mature.

More generally, it seems evident that growing shortages will turn the revolution of rising expectations into one of rising frustrations. This is already becoming apparent in many of the 'impossible' post-colonial creations, where the past decade has seen a yawning chasm between the hopes and programmes of the governments technically in power and the resource bases on which they have to operate. This discordance between what is conceived and what is conceivable in the political time scale is clearly shown in Guyana, which I shall cite simply because I know it better than others.

This is a country about the size of England, with a population little more than that of Liverpool, and of which 95% are clustered along a ten-mile wide coastal belt. Much of the interior is forest, difficult to penetrate as in Amazonia and continuing to present a hostile environment to migrants from the crowded coasts. However, in the south-western Rupununi savannas, forest at last gives way to grassland, and this is now luring the Guyanese to the deep interior just as the great temperate grasslands of the Prairies beckoned the cattlemen to the High Plains of the expanding United States. As in Brazil, roads are being built to the interior for the first time ever, designed to link up eventually with the road recently constructed north from Manaus on the Amazon to Boa Vista and Lethem.

The reasons underlying this Guyanese urge to open up the interior derive from the ethnic conflicts and resource constraints of an already fragile society. The East Indians, descendants of indentured labour imported after the abolition of negro slavery, are today the largest single ethnic group, and their high rate of natural increase is emphasizing their numerical superiority. This presents the African minority government with problems, met partly by what in Northern Ireland used to be called 'Derrymongering'. However, rather than having to rely on fictitious voters overseas, the government is now striving to import surplus negroes from Barbados—who are unable to emigrate to Great Britain of course— and settle these in the Rupununi as agricultural pioneers and loyal black voters.

The prospect of Guyanese politics being articulated permanently in such ethnic terms is itself sufficiently daunting; Black Power riots in neighbouring Trinidad and Tobago in 1970 stemmed from a similar pattern of racial cleavage, and such polarization seems bound to invoke repressive and authoritarian governments incapable of promoting stable economic growth. But quite apart from this prospect, the ecological outlook in the savannas is dismal. Despite the intoxicating sense of vast opportunities which distant horizons can give, these tropical grasslands are a seriously impoverished environment. Regular seasonal burning has been carried out for at least the 400 years that the Indians (or for clarity, the aboriginal Amerindians) have lived there, and the emaciated soil is now capable of carrying only about 30 head of cattle per square mile. With large investment in fertilizers and hydrological control it might be possible to increase this capacity substantially, but Guyana is quite unable to find the capital needed for such large scale and long term investment, and there is no certainty in any case that the net yields would justify any such allocation of scarce capital resources.

A further problem has already arisen from the improved economic and medical opportunities which transport developments since 1945 (particularly aircraft) have brought. The Amerindians of the Rupununi have begun to experience a population explosion as malaria eradication has taken place, and this is encouraging them to turn to animal husbandry as a means of livelihood. In other words, at the very time the government is seeking to send its favoured ethnic citizens to the savannas, these erstwhile 'empty' lands are rapidly filling up with a less politically trustworthy ethnic population. The scope of conflict is likely to grow, and here in a microcosm we can witness the group tensions which beset so many of the small succession states which have emerged in the aftermath of Western imperialism.

Unfortunately, too, the micro-conflicts are set in a global matrix of macro-conflicts. In the 1950s, the classic decade of the Cold War, it was customary to see newspaper maps of a monolithic communist empire straddling Eurasia, and bidding fair to pick off one by one the various promentories of the declining sea-based Empires.

During the 1960s this geopolitical nightmare began to recede, with the monolith splitting clean across the Asian heartland, and smaller cracks appearing in east-central Europe. Nevertheless, if this simple Stalinist model of the 'Two Camps' is no longer tenable, we still have the Maoist model of a rich and corrupt imperialist metropolis (including Moscow) confronting a vast 'peasant' hinterland of the world's exploited. We might argue that the patterns of confrontation are more complex than this, but it is impossible to wish away the sense of looming and portentous conflicts between the haves and the have-nots. If ecological crisis is indeed upon us, as I argue in my circulated paper, then there is all the more reason to anticipate a tense and dangerous 30, 40, 50, maybe 100 years, of conflict both at the micro- and the macro-level.

The sense of impending catastrophe, so succinctly summarized in the Forrester-Meadows graphs, is no doubt compounded of the traditional rich man's sense of guilt. The seismically plunging graphs are the numerate version of the Day of Judgement, when conspicuous consumption and profligate wastefulness shall be judged and found wanting spiritual virtues. No matter how we feed into the equations our technological pride and precocity, or our pollution controls and our perfect (sic) birth control, the Lord in His wisdom will make short shrift of all our pretensions. And great shall be the fall of all the graphs.

If the graphs are not lying, then we shall see thousands of millions of people dying in the space of some decades. Unless, that is, we can achieve Heaven on Earth, or more prosaically the stable-state society.

Perhaps because it is so easy to translate the doom-merchant's warnings into such hoary theological language, it is tempting to react with a no less old fashioned hedonism. No doubt this helps to explain the backlash against eco-catastrophe which we discussed earlier in this symposium. Yet the fact remains, and I am a great believer in the logic of facts, that in the space of a historical moment something has happened to our species for which there are no precedents whatsoever in any previous civilization. Nothing remotely approaching the recent increments in world population has ever happened before, and this exceptional period is far from having run its course. In my paper I have accepted the prospect of some ten billion human beings a century from now, yet even this would be substantially exceeded if rates of population growth remained at somewhere around their present 2% per annum. Extrapolating exponential growth at only 1% per annum shows that utterly absurd densities would be reached in a period of time equivalent to that which separates us from the beginning of the Industrial Revolution in Britain.

Now during this period when presumably the rest of the world will wish to industrialize too, resources will manifestly be consumed at an unprecedented rate. It is difficult to forecast precisely when certain key resources will be 'used up', if only because there are several variables involved. Thus the amount of uranium, what is termed 'potentially recoverable', will depend quite crucially upon changing prices (themselves a function of scarcity and demand) and changing technology (which would doubtless be stimulated by changing prices and the underlying shifts in relative scarcities).

Nevertheless, my paper rests on the premise that, quite literally, all the technological juggling in the world will not enable us to avoid a protracted period of growing scarcities in many, if not most, of the key material inputs of advanced industrial societies. I foresee these scarcities heightening political and military tensions between the rich world and the poor, and between the countries (or groupings) which form the rich world. It is well worth reading Hobson and Lenin on imperialism, even if we no longer have the latter's optimism about finding a political antidote. And it is also worth asking ourselves whether the sort of aggressive economic society which cheats or thrusts aside those, like the Amazonian Indian, who get in the way of progress, is not in the end going to devour all its vulnerable inhabitants. Which means all of us, in the end.

Discussion

RANDALL BAKER
In the case of the Organisation of Petroleum Exporting Countries (OPEC) group there is a major world need for the subject of the bargaining. What other commodities are there in which collective bargaining by Third World countries might be expected to be successful?

EDWIN BROOKS
Although OPEC is probably a special case, it is surprising how successful it has been, far more so than might have been predicted four or five years ago. Admittedly, at the present time, there are very few cases of commodities with bargaining potential. One possibility is rubber, where prices of synthetic rubber are on the increase, partly because of increases in oil prices. But I accept that OPEC is unlikely to be a precedent for similar action on most other commodities.

RANDALL BAKER
A point in relation to the OPEC negotiations is that they are helping one small part of the Third World, not being spread round equably.

EDWIN BROOKS
This is obviously a deficiency in the OPEC arrangements, and it is the height of irony that Iran is intending to use oil revenues, increased by the OPEC negotiations, to buy Concorde. Perhaps this is why we in this country are so amenable to OPEC, seeing it as just another form of subsidy for that confounded aircraft! There has been talk in OPEC of establishing a two-tier structure for oil with one price applying to developed countries and a much lower price applying to less developed countries. The idea

is pleasing but in practice it would be extremely difficult to operate.

RANDALL BAKER

I would suggest that a two-tier price structure is unlikely to come into operation for one reason alone, namely that the Middle Eastern oil producers are aware that their reserves are limited and they will want to get as much for their oil as possible. This will enable them to undertake the necessary internal and external investments sufficient to provide their countries with secure financial futures after their oil reserves have been utilized.

EDWIN BROOKS

That is certainly true and is one measure of the extraordinary amount of rethinking which has taken place in the past few years as the first signs of a major energy crisis have become visible. Consider the way in which the coal industry in this country, after being run down in the 1960s, is now being built up again. This is also likely to happen in the United States. Also, with reference to the United States, there appear to be negotiations in progress for that country to import large amounts of natural gas from the USSR's Siberian reserves. For the USA even to consider importing fuel from the USSR proves the magnitude of the crisis.

It is worth remembering that attempts to overcome fuel shortages by the use of new energy sources are fraught with difficulties. Nuclear power as an energy source involves new technologies which are certainly not fully tried. Furthermore, any switch to new energy sources has major social repercussions at the regional level involving running down of some industries with consequent unemployment and regional deterioration.

JOHN TYM

The more we talk about helping the developing countries, the more it seems to me that the answers to many of the problems lie with our policies and activities in the developed countries. Probably the best way to help them is to work to change the policies which we have adopted, whether these involve building massive motorway networks or new airports or other facets of a growth economy.

EDWIN BROOKS

I agree, but there is one important point to be remembered.
When seeking to change policies such as those you mentioned,
we do not have to appeal just to people's better nature. Many of
the allegedly economic arguments on which these projects are
based are themselves open to very cogent criticism. Take the
case of the Third London airport. If the criteria used in the
costing by the Roskill Commission were taken at face value,
then the best place for the airport would be Hyde Park! In fact
this exercise has already been carried out and published. Taking
the valuation which Roskill placed on the precious time of
business men together with that placed on historic buildings,
which the Commission related only to their insurance value,
then the case for Westminster airport becomes overwhelming.

Clearly there is something odd about this conclusion, just as
there are basic faults in the kinds of cost analysis so often used
as a basis for major projects. If attempts were made to achieve
more realistic cost benefit analyses, particularly taking into
account more of the income redistributive effects of these
projects, then the justification for many of them would be
greatly diminished. Of course, one would still come up against
all the vested interests which would make change difficult. A lot
of people make a lot of money out of the motorway building
programme.

MILES DANBY

I don't think we should concentrate solely on the large projects
such as Concorde and urban motorways, but should also pay
attention to the ways of conserving energy by the use of low
energy methods in construction and use of buildings. There are
many ways in which this can be done, and research in this area
deserves all the encouragement possible. Above all this involves
positive action rather than negative criticism.

JOHN TYM

While that is certainly desirable, we have to continue the critical
approach. Otherwise there is a danger that although the kinds of
activities Professor Danby wants may take place, we will also
continue to get the large projects. This is particularly so because
of the power and influence of the vested interests concerned
with promoting the large projects.

EDWIN BROOKS
I accept both points, the need for strong criticism in some areas together with more positive approaches in others. One other point is that it is going to be a very long process; it would be too hopeful to expect any short sharp campaign bringing effective long-term results.

One of the problems in this kind of campaigning is that the conservation movement has tended to be regarded as a lot of hair-shirted back-to-the-land individuals, not to be taken too seriously. It is, in fact, a movement concerned with the basic structure of society and it must dispel this image of remoteness from the real world of ordinary people.

Workshop Discussion

During the symposium, workshops were organized to enable the participants to discuss the subject matter of the symposium in small informal groups. To assist discussion, the six groups were asked to comment on a paper entitled 'Environmental Considerations for Development in the Third World'. This paper, which is included as an appendix to this volume, was produced by a working party of the World Development Movement.

WDM is an organization concerned with international development and made up of a large number of local groups situated throughout Britain. The paper was accepted by the WDM Council on 10 March 1973 as a statement of its position and a basis for discussion with environmental organizations. It was made available to the symposium organizers by the Secretary of the WDM, Sarah Wells. The following summary of the workshop discussion gives the main points raised during that discussion.

1 Resource Utilization

There was repeated emphasis of the need for a very considerable reduction in the rate of use of non-renewable biological and physical resources by developed countries. Low impact technology has, in the past, been considered primarily with reference to less developed countries. As part of a trend towards resource conservation, it must now receive far greater emphasis in the context of developed countries.

In resource exploitation, one one of the major problems will lie in the field of resource appraisal. As pressures on resource use increase, there will be an increase in political and economic pressure exerted by developed countries and by multi-national corporations on less developed countries to make them exploit their resources quickly. In this context, the possibility arises that developed countries and multi-national corporations will be able to obtain, by means of satellite reconnaissance, information on the resources of less developed countries which is not available to the governments of those countries. An international mechanism is needed to prevent this kind of abuse.

2 Research and Development

Concerning both development strategy formulation and the utilization of research findings in Third World countries, a persistent danger is that of exporting attitudes from developed countries which are irrelevant to the situation in less developed countries.

There is a parallel danger that the attitudes of some nationals of less developed countries are excessively influenced by their contact with developed countries and give rise to similar effects. One of the means of overcoming these problems is by increasing the applied research capacity located in and controlled by less developed countries.

A specific problem concerning applied research relates to the ecological effects of development programmes. Any research on these effects must necessarily be long term, studies of the order of ten years or more in length being required.

3 Industrial Development

There is a tendency for industrial development to be fostered in less developed countries via private investment by commercial interests in developed countries. Apart from the fact that this may not be in the best economic interests of less developed countries, there is the very real danger that such development will be environmentally and socially damaging. A particularly retrograde tendency lies in the placement of industries in less developed countries which are considered too dirty for location in the developed industrialized countries in which they originated. It is desirable that less developed countries should appraise themselves fully of the problems of industrial development which have been encountered in developed countries in order to avoid these tendencies.

4 Aid

While recognizing the need for increase in aid, emphasis was given to improving its quality. Multilateral agencies were considered preferable to unilateral ones, but more attention needed to be placed on ensuring that projects were relevant to the stated needs of the less developed countries concerned. Long term projects with sound ecological pre- and post-auditing were advisable.

5 Education

In both developed and less developed countries, an important educational priority is the union of environmental and developmental approaches at all levels of education. This requires a large measure of interdisciplinary education.

6 Political Realities in Developed Countries

Limitations of resource utilization reinforce the need to achieve that global redistribution of wealth necessary for the long term management of the planet. In the present climate of opinion, in developed countries, it is extremely unlikely that such a redistribution will ever be seriously considered. In such a context, how can we best seek to achieve the necessary change in that climate of opinion? This is the central problem for environmental and development interests in developed countries.

It must be approached realistically and we have to recognize the need to achieve fundamental changes in our approaches to education, much of which is irrelevant if not positively dangerous at present. One of the few positive signs is that there has been a discernible increase in the interest of people in their local environment—can they be educated to care for their global environment as well?

General Discussion

KENNETH DAHLBERG
Some stress has been laid on the importance of team work in undertaking research and development. This is seen to be important in ensuring that some kind of synthesis is achieved in the approach to a project, thus making it more likely for that project to be successful.

I would suggest that the team approach is not necessarily the way to achieve a synthesis of the approaches of different disciplines in that this can best be achieved on an individual basis. It is important that anyone involved in development programmes should attempt to 'de-specialize' sufficiently to enable him to grasp the approaches of the disciplines which impinge on his work.

There is an analogy here with learning a language in that you have to learn not only the vocabulary and grammar but also something of the culture as well, enough to be able to translate the spirit of things as well as the purely mechanical translation. Our most sophisticated technology is rarely able to translate directly into different cultures and environments.

We have touched on matters concerning education but one point which I would stress is the need for re-education among ourselves, particularly in the area of appreciating the workings of our total environment. We need to become aware of the environment of which we are part and to understand something of its workings. This is, I think, a matter of central importance, and one which is barely under way.

JONATHAN HOLLIMAN
The process of development has been dominated for far too long by the idea that it is a linear process, for example a progression from subsistence agriculture through to intensive beef production. From an ecological point of view it just

doesn't work like that, and I doubt whether it works from many social points of view. We are too concerned to measure development purely in terms of growth in GNP when there are many other parameters which have to be taken into account as well. Is human happiness really related just to wealth? I certainly don't see how it can be.

In defining what is meant by development we have got to define what is meant by environmentally sound development, what is meant by appropriate technology and what can be determined concerning human happiness in relation to development aims. All these other criteria must be taken into account.

SARAH WELLS

The papers which we have heard have given an indication of current progress in world development and have also identified the major environmental problems involved in world development. My general feeling is one of guarded optimism in that I think there is still time to take into account these newly realized considerations in future world development plans. What is important is that this symposium should not be seen as an end in itself but rather as an opening stage in the re-orientation of our priorities and outlook concerning development.

In practice we have to stimulate the development of the type of policies which recognize environmental considerations, and we have to do this by whatever means are available to us. This might involve encouraging work on appropriate technology, or work on predictive modelling. Whatever the faults of the 'Limits of Growth' approach, it does present us with possible options based on attempts to predict future change resulting from current policies. The alternatives need to be presented to the decision makers—politicians and public—so that conscious choices can be made.

KATHRYN MORTON

From the point of view of an economist, I would say that much of the criticism of economists which comes from environmentalists is misplaced. In criticizing economists, they are usually criticizing values and outlooks which society as a whole has endorsed. The danger in this is that economists can get a bad name to the extent that their views and particularly their methods are not utilized in development planning. This is a considerable loss because economists have a great deal to contribute to any attempts to undertake broadly based development studies.

PETER KENYON

Perhaps inevitably, a divergence in approach is evident between the ecologists and the economists on matters of development. In one sense they have similar outlooks. The ecologist, looking at developed countries, sees the dangerous over-exploitation of resources which goes on. The economist concerned with less developed countries, is aware of how the exploitation of these countries continues. One sees the waste of human resources, the other the waste of material resources.

Where there is a degree of conflict is on the question of time scales. Here, the economist sees many problems associated with development but anticipates the possibility of achieving a measure of development in the long term. The ecologist, on the other hand, is concerned about limits to total world growth and the possibility that such limits will seriously hinder Third World development, thus frustrating the hopes of the economist. The economist sees continual growth as the only way of ensuring a future for less developed countries, but the ecologist says that this will not be possible.

Even so, I am optimistic about the chances of the two approaches coming together. There has already been a discernible change of emphasis among development economists away from growth as the basic requirement in development. It is more important to encourage this than to spend one's time hitting out at economists in general. Ecologists are demanding a major restructuring of our society and they need economists as allies if there is to be any chance of success.

BRUCE DINWIDDY

There is a growing realization that the development strategies of the majority of less developed countries are wrong, and they are wrong both through our fault and through the fault of the countries themselves. The processes of development we follow are ecologically irresponsible, and less developed countries are adopting similar processes.

The job of changing development strategies to take into
account ecological perspectives starts in the developed countries
and it is a major political activity. At the same time, we all have
to recognize that as individuals in developed countries we may
be experiencing a standard of living which would be incompatible with ecological realities in a world in which wealth was
equally distributed. The recognition of this personal
predicament is an essential prelude to changing attitudes in the
country as a whole.

G S PURI
It seems especially important that we make an attempt to
establish the central place of the idea that, in the final analysis,
we are all dependent on each other.

EDWIN BROOKS
Although that may seem wildly optimistic, there are examples
from history of high degrees of community loyalty and
awareness having been achieved. I am thinking, for example of
the record of the nation states, and in particular of the
community awareness they succeeded in generating in times of
war. It has been said that Stalin won the Second World War
when he finally called on Russian nationalism.

There do exist, on occasions, powerful forces which
encourage people to identify with other human beings for a
common purpose. The problem of course is that on previous
occasions other human groups were seen as the enemy
threatening the community. But at this present time the threat
is one which affects us all, the threat of an environmental crisis.
The important thing therefore is to articulate this threat in such
a way that the world community is shocked into a sense of
common danger which needs to be met by common sacrifice.
Such a new awareness would involve major changes in the
policies of the developed countries, notably the development of
resource-conserving economies and a redistribution of wealth
towards the less developed countries.

The obvious difficulties suggest that this new awareness will
not come overnight, and it may need decades or even
generations to make people fully aware of the fundamental
changes which growing scarcities will force upon them. But
there are some hopeful signs, such as the degree to which

younger people today are totally disinterested in—and even positively opposed to—old-fashioned nationalism. The question is whether such a mood can lead on to community awareness on a global level; if not, then the coming period of increasing resource shortage is likely to see a reassertion of old tribal hostilities, with all the dangers that this would mean.

G S PURI
The most impressive example, in terms of numbers involved, of post-war community awareness has been Mao's China. Do you think that the Chinese example could be repeated in developed countries or other less developed countries?

EDWIN BROOKS
There is no doubt that the most successful movement of the twentieth century has been communism, although the local versions have varied. This is a movement which does at least pay lip service to the idea of world brotherhood united against a common enemy of exploitation. It is a movement which has left behind the most appalling carnage. but this tends to make us forget the degree to which it has so often succeeded in uniting large populations in a common cause.

Mao's power base was built up with the peasant as its core, and although the Chinese model may not hold good for developed industrialized countries, in the less developed world it may be a model relevant to more countries than we care to admit. We are now beginning to appreciate that future environmental pressures limiting human activity may have their greatest impact in the field of agriculture. In that the Chinese model is the form of communism which has the peasant farmer as its base, then that model must be taken seriously in the context of the environmental crisis.

ROBERT DICKSON
Most discussion about different forms and models of development misses the point, which is the problem of underdevelopment and its causes. This is central to any attempt to introduce effective means of development. The prerequisite is to understand thoroughly the causes of the phenomenon of underdevelopment, the situation where a society is prevented from taking the decisions which will ensure its development.

Third World countries are prevented, by the domination of the developed countries, from undertaking the programmes which will result in their development.

DAVID COTTON

Difficulties associated with changing educational systems are also at the root of many problems of underdevelopment. Many less developed countries have academic educational systems in their schools which are totally out of tune with their present needs; systems which remain as legacies of previous colonial rule. It is an extremely long process to induce improvements because the people continue to regard an academic education as being a prerequisite for administrative appointments. It is also their experience that practical skills rarely enable a person to obtain a standard of living comparable to that enjoyed by people who have managed to successfully enter the administrative system. Consequently the population still demands an academic education and there is tremendous resistance to the introduction of vocational training and other non-academic forms of instruction which are probably more relevant to the country's present needs.

BRUCE DINWIDDY

Change in less developed countries' educational systems depends on appropriate political leadership. The key problem, though, is that as long as you have an educational system with a distinct academic core, then that aspect will remain the most sought after and anyone who does not proceed to take academic subjects will be regarded as inferior. Basic vocational training for everyone may be the one essential in less developed countries as an aid to development. Any selection of people for more advanced education has somehow to avoid giving that aspect the appearance of elitism. This is, to say the least, very difficult to avoid.

EDWIN BROOKS—Final Comments

It would be impossible to sum up this symposium, especially as it is not the sort of conference for uttering grandiloquent slogans from which we all derive virtuous good feeling! We have recognized that we are dealing with a very complicated range of problems, problems of such a magnitude that they may constitute the most serious crisis yet faced by the human race. Understandably, we have done no more than touched on these problems, but we hope that our meeting will contribute towards bringing together the common interests of those concerned both with development and with the environment.

Postscript

One of the aims of the symposium on which this book is based was to bring into focus the relationship between the subject matter of the two major UN conferences held in 1972. The third session of the UN Conference on Trade and Development (UNCTAD 3) held in Santiago and the UN Conference on the Human Environment held in Stockholm each attracted interest in this country, although Press activity on the Stockholm conference far exceeded that on UNCTAD 3. Little attempt was made to link the conferences together, even though each was concerned with the state of the human environment.

During the course of the symposium, three attitudes, not necessarily mutually exclusive, became evident and these tended to reflect dominant national attitudes at the international conferences the previous year. There were two approaches to solving the problem of world poverty. According to one, the only hope for Third World development lay in self-determination by the countries concerned. By pursuing economic self-reliance on the Chinese or perhaps Tanzanian model, they would be able to control their own future development and would not be subject to the control currently exerted by the developed countries which exploit them. A second approach was concerned with attempts by less developed countries to exert more influence on the world trading system, which, combined with better commodity production planning, would ensure for them a considerable growth in revenue from exports.

A number of ecologically inclined participants were more concerned with the effects of more basic limitations to total world development. Palmer Newbould expressed this towards the end of his paper when he discussed problems likely to arise from future pressures on the biosphere resulting from increased population size combined with increased personal aspirations. He was of the opinion that, whatever the success of population control programmes the world's population would at least treble in size before stability could be achieved. If the expectations of nearly 12 billion people were in line with current resource use in, for example, the USA, the demand on world resources would increase approximately fifteen-fold. In his opinion, the global ecological carrying capacity would then be seriously exceeded. In other words, the world could not support such activity.

Putting this in the context of world poverty and the means of alleviating it, he foresaw global constraints on development set by resource and environmental factors. At some point in the near future, therefore, resource use by less developed countries would need to be accompanied by a *decrease* in resource use by developed countries, 'a levelling down as well as up' as he put it.

Other speakers elaborated on aspects of this theme and it was left to Edwin Brooks in the concluding paper to look at the implications of this view. His three postulated scenarios were, first, the rich world stripped of its privileges and wealth by the bargaining strength of the developing countries, secondly, the reverse picture with the rich countries using their power to preserve their privileged position, and finally a triumph of liberal attitudes involving resource conservation and an end to economic growth by the rich, greatly increased help for the poor and far more effective population control on a global basis, all stemming from the common recognition of the need to conserve the planet.

Central to all this is the idea that global stability and peace can only be achieved by action to prevent world poverty in which the rich must give up a part of their wealth. This apparently idealistic notion is, according to some participants, not a matter of idealism at all but of ecological reality. It runs counter to the more traditional approach to development with the Third World getting rich only as part of a continual increase in world wealth. Their slice of the cake may not grow but the cake does grow so their wealth slowly increases. According to the ecological view, the cake cannot go on growing forever and the poor will only rid themselves of their poverty by increasing their share of the cake at the expense of the rich.

If this is correct, then the implications are immense. The symposium did no more than touch on them, but if, by this book, more people can begin to appreciate the implications, then its publication will have been of value.

Appendix I. Environmental Considerations for Development in the Third World

This paper was accepted by the World Development Movement Council on March 10th 1973 as a statement of its position and a basis for discussion with Environmental Organizations. It is not a final definitive statement and may be subject to amendment and amplification.

A. Position of the World Development Movement

Preface: The World Development Movement believes that all human beings should have the opportunity to develop their physical, mental and spiritual potential to the full without the restraint of hunger, poverty, disease and illiteracy and without physical oppression or political suppression. We believe that the present situation, in which the earth's resources are controlled and squandered by the few, is fundamentally unjust, is an indictment of the values of our current Western civilization and poses a threat to future generations.

1. Introduction—Population, Resources and Unemployment

The consumption of resources by people in the Third World is inadequate to raise the standard of living above the poverty level. The immediate causes of the low consumption of the poor are: low returns for work done and massive unemployment. A complex set of technical and socio-economic factors, local and international, account for the low productivity and returns. Great care must be taken to ensure that solutions to these problems do not increase unemployment. The need to raise levels of employment is a basic requirement of any policy affecting developing countries, including policies relating to technology. Technology also concerns Environmentalists from the point of view of resource consumption and pollution. The aim of both Environmental and Development Lobbies must be to increase consumption per head in poor countries, both by raising their own productivity and by effecting a transfer of resources from the rich world by appropriate aid and trade policies.

2. Population

In many Third World countries population is growing so rapidly that increases in food production, jobs and goods are being absorbed by population growth without improving the standard of living. A reduction in the rate of population growth would buy time for Governments to achieve real improvements in the standard of living of their people. At the same time no form of family planning stands any chance of reducing the growth rate unless accompanied by a real increase in the standard of economic well being and social justice. Both Environmental and Development Lobbies, therefore, need to recognize that the goals of reducing population growth and accelerating economic development in the Third World are mutually dependent and that they need to press simultaneously for both types of policies.

3. Resources—Recycling

Recycling of all scarce resources from agricultural wastes in sewage to scarce metals is ultimately inevitable. Indeed, craftsmen in poor countries are already skilled at recycling in a small way, used tins and rubber tyres into useful household articles. As certain resources become depleted the mineral reserves of many developing countries will increase in value in the long term but restrictions on the use of these minerals could affect their short term balance of payments adversely by reducing demand in relation to supply. Two lines of action are required: international control of the mineral markets on the one hand and, on the other, aid and trade policies geared towards developing other sectors of the economies of countries at present heavily dependent on mineral revenues.

4. Pollution—Environmental Tax

The WDM is deeply concerned about various proposals relating to types of environmental tax such as taxing polluting industries or placing import restrictions on food grown with the help of harmful pesticides. The developing countries would be at a considerable disadvantage if they had to meet the additional costs involved in starting new industries of their own and might have to pay even higher costs than they do now for imported industrial equipment while the rich, with their already established industries, would be able to pay the penalties and continue production. The granting of pollution tax exemption to developing countries could make them a tax-free haven for polluting industries and, while some developing countries might welcome this where local pollution is not yet appreciable, it could only be a short term solution and could not solve the global problems. Environment and Development Lobbies should encourage the development of non-polluting technologies but at the same time ensure that proposals do not hinder the development efforts of the Third World.

5. Rural and Urban Development

Rural development should have priority, first because most people in developing countries still live in rural areas dependent on the land and, secondly, because towns in the Third World are expanding faster than jobs and housing can be provided for the urban immigrants who consequently live in conditions worse than those they have left in the villages. The aim should be to raise productivity of the farmers; provide jobs for those the land cannot support; improve living standards on the spot by means of village development projects such as health services, water supplies, sanitation (including methods of recycling sewage); education and village industries. At the same time life in the cities must be improved by expanding industrial employment and carrying out environmental improvements in terms of housing, transport, sanitation and water supplies.

6. Agricultural Productivity—Pesticides and Fertilizers

If food production is to increase in the Third World in the immediate future, all known technology will need to be applied including increased amounts of fertilizers and pesticides, both little used at present in the developing countries. DDT, the major organo-chlorine compound pesticide, has been invaluable in public health programmes and in reducing crop losses. But in view of the growing evidence of the harmful effects of DDT on fish and birds, it should be substituted by better methods as soon as possible. Alternatives to DDT are more expensive and have hazards of their own. The environmentalists and development lobbies together should press for the means to be made available to meet additional costs of sound methods of pest control and for improving agriculture along environmentally as well as economically and socially sound lines.

7. Technology

In most developing countries the ratio of the factors of production, i.e. scarce capital and abundant labour, combined with low levels of education and skills, means that intermediate technology will constitute the most rational use of resources. This does not preclude, however, the use of the most up-to-date and sophisticated technology where appropriate to development needs.

8. Trade or Self Sufficiency

WDM rejects any suggestion of a system of nationally self sufficient states and insists on the value of international trade for economic, pragmatic and ethical reasons. Economically, trade can make possible the most efficient use of economic resources to the advantage of all participants. Pragmatically, an interdependent world society has a greater incentive for peaceful co-existence than independently self sufficient societies. Ethically, the pursuit of self sufficiency denies the sharing of our world's resources for the mutual benefit of all.

B. In the Interests of Third World Development and the Safeguarding of the Global Environment the Following Policies are Urged Upon the Government

AID

1. An increase in the total size of the aid programme is required to effect a substantial net transfer of resources from rich to poor countries.

2. A reorientation of aid is required to achieve environmentally sound economic and social development for the masses of the people in developing countries:

(a) Reduce the proportion of aid: (i) given for import-requiring projects; (ii) tied to purchases of equipment imported from the donor.

(b) Increase the proportion of aid given to development projects which: (i) are labour intensive; (ii) have high local costs; (iii) use appropriate technology; (iv) are environmentally sound; (v) assist diversification of economies at present dependent on mineral resources threatened by exhaustion and international conservation measures.

3. Provide aid in addition to the internationally accepted level to cover the cost of the use of more expensive but safer pesticides.

4. Provide resources (technical, manpower and funds) for:

(a) Research into development of non-polluting technologies appropriate for developing countries.

(b) Research into the development of ecologically sound agricultural methods including the application of fertilizers and pest control (e.g. expansion of the Tropical Products Institute).

(c) Promotion of these technologies and methods through education and extension work.

5. Provide aid for family planning programmes for those countries which wish to receive it.

Trade

1. Promote and join international agreements which safeguard the export earnings of developing countries.

2. Establish more generous quotas and reduce tariffs on primary and manufactured products from developing countries.

Private Investment

1. Provide assistance to developing countries:

(a) To assess the economic, social and environmental effects of proposed investments.

(b) To draw up, monitor and enforce appropriate agreements with investing companies.

2. Ensure that private investment does benefit the developing country by the negative use of official incentive schemes: e.g. the investment insurance scheme and the schemes to use official aid to promote investment, should apply to investments in *only* those developing countries which operate an effective assessment procedure.

Appendix II.
Symposium of Participants

Paul Arnold	Selby, Yorkshire
Mrs C Arnold	Selby, Yorkshire
Dr R M Auty	Simon Fraser University, British Columbia, Canada
Dr Randall Baker	School of Development Studies, University of East Anglia
Miss Lynda Barber	University College, Cardiff
Dr Kenneth Barlow	The Old Forge, Great Finborough Suffolk
R Barrowclough	Department of Geography and Geology, The Polytechnic, Huddersfield
Mrs J P Bolam	Alsager College of Education, Crewe, Cheshire
Dr David Brady	Department of Applied Chemical and Biological Sciences, The Polytechnic. Huddersfield
Shelagh M Brooke	Department of Applied Chemical and Biological Sciences, The Polytechnic, Huddersfield
Dr Edwin Brooks	Department of Geography, University of Liverpool
Dr B J H Brown	Department of Geography and Geology, The Polytechnic, Huddersfield
Roland Chaplain	Huddersfield Technical College
Wendy Cole	Department of Applied Chemical and Biological Sciences, The Polytechnic, Huddersfield
Dr David Cotton	School of Environmental Sciences, University of Bradford
Professor Kenneth D. Dahlberg	Department of Political Science, Western Michigan University Kalamazoo, Michigan, USA

Professor Miles Danby	School of Architecture, University of Newcastle, Newcastle upon Tyne
Mrs Zena Daysh	Commonwealth Human Ecology Council, 63 Cromwell Road, London SW7
R Dickson	World Development Movement (Northern Office), Leeds, Yorkshire
Jean Dixon	University of Bradford, Yorkshire
Bruce Dinwiddy	Overseas Development Institute London W1P 0JB
Paul Downton	University College, Cardiff
L R Foweather	Barnsley, Yorkshire
J H Fox	Ilkley, Yorkshire
J A Grace	Bretton Hall College of Education, Barnsley, Yorkshire
James Grant	Portincaple, Dumbartonshire, Scotland
Dr Leslie Hale	Centre for Human Ecology, University of Edinburgh
Mrs D Hall	Department of Education, The Polytechnic, Huddersfield
Dr David Hall	Overseas Development Administration, London SW1
Mrs E Hims	Leeds, Yorkshire
J Holliman	Friends of the Earth, London
Angus Hone	Institute of Commonwealth Studies, Oxford
Miss C Johnson	Leeds, Yorkshire
J Owen Jones	Commonwealth Bureau of Agricultural Economics, Oxford
Peter Kenyon	54 Devonshire Road, London SE23 3SX
Dr David Kirby	Department of Geography and Geology, The Polytechnic, Huddersfield
J D Lawrence	Keighley, Yorkshire
Dr R Livingstone	Department of Applied Chemical and Biological Sciences, The Polytechnic, Huddersfield
D K McIntyre	Halifax, Yorkshire
Kathryn Moreton	Overseas Development Institute. London W1P 0JB
Professor P J Newbould	New University of Ulster, Coleraine, Northern Ireland

Jan Newton-Howes	Department of Applied Chemical and Biological Sciences, The Polytechnic, Huddersfield
Miss H M Nutton	Almondbury Huddersfield
A O'Driscoll	Gomersal, Cleckheaton, Yorkshire
Rhys Owen	University College, Cardiff
Kenneth Parry	81 Onslow Square, London SW7
David Pickles	University College, Cardiff
Alan W R Potter	Department of Geography and Geology, The Polytechnic, Huddersfield
Dr G S Puri	Department of Biology, Liverpool Polytechnic
Susi J Reeves	Airedale and Wharfedale College, Pudsey, Yorkshire
Dr H Robinson	Department of Geography and Geology, The Polytechnic, Huddersfield
Dr Paul Rogers	Department of Applied Chemical and Biological Sciences, The Polytechnic, Huddersfield
Clive Simms	Bretton Hall College of Education, Barnsley, Yorkshire
Sir Hugh Springer	Association of Commonwealth Universities, London WC1H 0PF
E W Stephenson	Wakefield, Yorkshire
J L Summerscales	Stairport, Barnsley, Yorkshire
B Swindell	Larks Hill, Pontefract, Yorkshire
Anthony Tucker	'The Guardian', Grays Inn Road, London WC1
John F Tym	Department of Modern Arts, Sheffield Polytechnic, Sheffield. Yorkshire
Anthony R Vann	Department of Applied Chemical and Biological Sciences, The Polytechnic, Huddersfield
Stuart Walton	Hoylandswaine, near Sheffield, Yorkshire
Sarah Wells	Secretary, World Development Movement, Department of Geography, Portsmouth Polytechnic, Portsmouth, Hampshire.

Appendix III.
Notes on Contributors

EDWIN BROOKS

Dr Edwin Brooks is Senior Lecturer in the Department of Geography at the University of Liverpool and was, from 1964 to 1970, a member of parliament. He is a Past President of the Conservation Society. He is the author of a recently published book on population problems and policies in Britain entitled *'This Crowded Kingdom'* and recently spent some time in South America studying development problems.

KENNETH A DAHLBERG

Professor Kenneth Dahlberg is a member of the staff of the Department of Political Science at Western Michigan University, USA. He recently spent a year at the Institute for the Study of International Organisations at the University of Sussex and it was during this sabbatical year that he participated in the symposium. While at Sussex, his research interests were mainly concerned with the 'green revolution' especially the social and political effects of introducing new agricultural techniques and processes on a massive scale.

BRUCE DINWIDDY

Bruce Dinwiddy was until recently a research officer at the Overseas Development Institute (ODI) in London. After reading PPE at Oxford, he was awarded an ODI Fellowship under which he worked for two years as an economist in the service of the

Swaziland Government. He joined ODI's research staff in 1970, and has been mainly engaged on a study of indigenous entrepreneurship in Africa, with a view to identifying ways in which non-agricultural entrepreneurship may be stimulated or assisted through external aid programmes. He edited ODI's *Review 5* (1972) and *Review 6* (1973) which examined the overseas development policies of Britain and other rich countries.

JONATHAN HOLLIMAN

Jonathan Holliman studied geography and botany at University College, London. He has worked on many environmental programmes, both in Britain and in several non-industrialized countries. For the past three years he has been one of the directors of Friends of the Earth and now undertakes freelance research and writing. He is interested in the relationship between environmental problems and the development process and is author of '*The Ecology of World Development*' a booklet published by VCOAD, London, at the time of the Stockholm environment conference.

ANGUS HONE

Angus Hone is a research officer in the Economics of Development at the Institute of Commonwealth Studies, Oxford. He worked from 1968 as an officer of the International Trade Centre (GATT/UNCTAD), Geneva, spending the bulk of each year working in Asia as an adviser to governments on problems of trade. He is an economist principally interested in South Asian development problems.

PETER KENYON

Peter Kenyon took a degree in Economics at the university of Hull and then went to the University of Leeds where he studied for an MA in Economics of Development. He recently completed an 18 month period as Research and Information Officer for the Voluntary Committee on Overseas Aid and Development, during which time he wrote on a wide range of development topics.

PALMER NEWBOULD

Professor Newbould is Professor of Biology at the New University of Ulster. From 1955 to 1967 he was Lecturer in Plant Ecology at University College, London and was in charge of the Diploma and MSc course in Conservation there, the first course of its kind in Europe. He is a Council member and Vice-President of the British Ecological Society and is involved in the work of the Nature Conservancy and the International Biological Programme. His main research interests are in primary production of ecosystems and conservation.

PAUL ROGERS

Paul Rogers is a lecturer in the Biology Section at Huddersfield Polytechnic where he is mainly concerned with courses on human ecology. From 1968 to 1970 he worked on a regional agricultural development programme in East Africa and more recently spent a short time in Hong Kong. He is a member of the Executive Committee of the Commonwealth Human Ecology Council and edited the proceedings of the 1972 human ecology symposium at Huddersfield entitled 'The Education of Human Ecologists'.

SIR HUGH SPRINGER

Sir Hugh Springer, who chaired the first session of the symposium and wrote the Foreword to this volume, is the Secretary-General of the Association of Commonwealth

Universities. He is also Chairman of the Commonwealth Human Ecology Council.

ANTHONY TUCKER

Anthony Tucker is the Science Correspondent of 'The Guardian' newspaper. Although his work necessarily covers a wide range of subjects, two of his main interests are in the fields of energy resources and heavy metal contamination. He has contributed a number of major articles on energy resource utilization to 'The Guardian' and is author of a book called 'The Toxic Metals' published in 1972. This was probably the first well-documented but popular account of the extent of environmental contamination by heavy metals.

ANTHONY VANN

Anthony Vann took his first degree at the University College of Wales at Aberystwyth and followed this with three years research on grassland ecology, also at Aberystwyth. Although his main teaching and research interests are in plant ecology, he has become increasingly involved in courses in human ecology since taking up an appointment at Huddersfield Polytechnic two years ago.

Subject Index

Author Index